A Look at
The Power Supply

# A Look at The Power Supply


Stephen Hodson

B.Ed. (Hons.) SDES

*Assistant Director, ESTED, Clarendon Laboratory, Oxford*
*(European Science & Technology Database Project)*

*(Formerly Lecturer / Consultant, NEMEC)*


Edited by
David Samuel Charles Thompson


NEMEC
The University of Southampton
SOUTHAMPTON
SO9 5NH
England

ISBN 1 85354 060 9

Published by NEMEC, The University of Southampton,
Southampton, SO9 5NH, England.
Telephone: (0703) 558379.



Typeset by the author and the editor.

Printed by Hobbs the Printers Ltd., Southampton.

**British Library Cataloguing in Publication Data**

Hodson, Stephen, *1947-*
A look at the power supply.
1. Electronic equipment. Power supply devices.
Design & construction
I. Title
621.381044

ISBN 1-85354-060-9

ISBN 1 85354 060 9

*To Fiona and Alexander*

# CONTENTS

# PREFACE

Soon after the last World War, which he spent maintaining radio equipment for the benefit of the 8th Army, my father presented my mother with two gifts which were to eventually result in the production of this monograph. One of them was me, the other was ten tons of redundant radar equipment. The latter 'present' meant that my formative years were spent surrounded by the paraphernalia of electronics. Strangely, I have only become a licensed radio amateur (as G1XTA) in recent years, my earlier interest being in the construction and flying of radio controlled model aircraft and their radio equipment.

Following four years at Leicester University I spent some sixteen years teaching physics, electronics and computing before joining NEMEC. This monograph has its roots throughout this history. Dissatisfaction with the texts available for use in schools always led me to produce my own written materials, especially for Advanced Level students taking optional electronics papers, and while preparing material for the advanced courses offered by NEMEC it became obvious that teachers have a need for texts which treat electronics at a greater depth than that usually met with in school textbooks, without such complex mathematical analysis as is normally found in undergraduate texts.

The need for a text about power supply systems in particular was reinforced when a review of the available books, and articles in magazines, revealed that some old myths about the behaviour of power supplies were still extant. In particular the traces obtained by connecting an oscilloscope to the rectifier stage of a smoothed power supply seemed to bear little relation to those in the textbooks. It is hoped that in this text the reader will recognise what he sees in practice. Further, as I talked to teachers attending courses at NEMEC, it became obvious that much poor and potentially dangerous practice still existed in some schools, where pupils were building and investigating open mains power supplies. It is hoped that the methods outlined in this text will go some way to preserving both teachers and their pupils. This book should also serve the needs of the radio amateur/hobbyist who needs a power supply for a particular application but does not wish to commence the design from first principles.

In the following pages you will find an introduction to the subject of power supplies. A systems approach has been adopted so that the reader is carried gently forward into a steadily increasing complexity of subject matter. Sufficient mathematical background has been included to meet the needs of the student of advanced level physics, electronics and technology who is assumed to be competent in manipulating the formulae associated with Ohm's Law. Wherever possible, circuit diagrams are presented, which may be constructed with a fair certainty that they will function as desired, culminating in the outline of how one project, exclusively concerned with power supply systems, was approached.

I would like to thank my colleagues David, Mike, and Patricia for their guidance and encouragement in the production of this text.

Stephen Hodson

November, 1989

# THE FUNCTION OF POWER SUPPLIES

The function of a power supply is to take an electric supply and modify or control it in order to supply other electronic systems with the correct levels of voltage and current which enable them to operate correctly. The power supply should also provide protection, both for the user and the equipment, in the event of a fault developing within itself or the system being supplied.

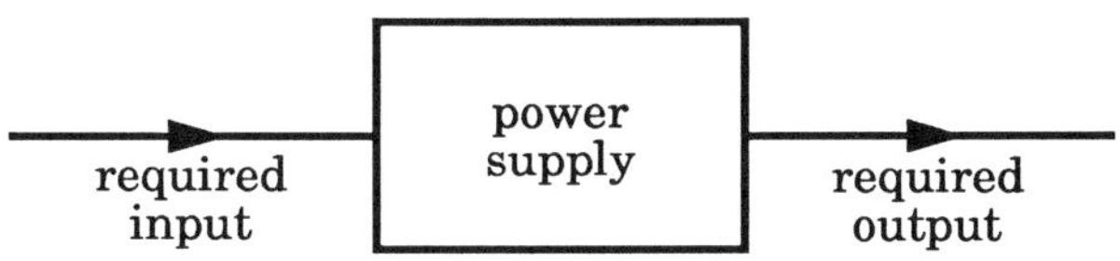


**Figure 1. The power supply system.**

The sources of electrical power normally available for use by electronic equipment are the domestic alternating current supply and batteries.

The domestic electricity mains supply is a single phase alternating potential difference with, nominally, the characteristics of a sinusoid, frequency 50 Hz and p.d. 240 $V_{rms}$. An examination of the supply voltage will often reveal the presence of 'spikes', transient short pulses which have amplitudes perhaps in excess of 100 V, and other signals superimposed on the supply. Further, the frequency may vary by a small percentage and the amplitude may vary by up to 6% especially during periods of high load. The function of the power supply is to convert the supply from a source of alternating current to a source of direct current suitable for the equipment supplied.

The lethal nature of the mains supply has resulted in several examining boards specifically proscribing the presentation of mains powered projects for assessment and the feeling of the author is that the use of open equipment carrying mains potentials by pupils is an unnecessary hazard not to be undertaken in schools. However, school examination syllabuses all require knowledge and understanding of the implementation of power supply circuitry. There is no reason why work cannot safely be undertaken using low voltage alternating current supplies derived from a suitably protected, commercially available, supply.

Batteries have the intrinsic drawbacks of high cost, limited current availability, increase of impedance with age and drop of terminal potential difference (p.d.) as they discharge. The employment of batteries may therefore necessitate the use of control circuitry with the functions of stabilising the output from the supply and reducing the impedance 'seen' by the circuitry being supplied. These functions and the means of implementing them will be included in the discussion of the control of mains derived power supplies.

The functions of the mains operated power supply will be considered as twofold. Firstly the conversion of the transformed alternating current to a direct current and secondly the supply of this direct current to required specifications.

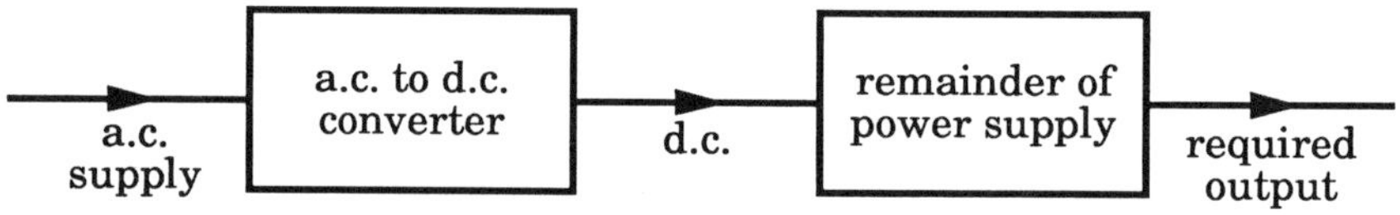


**Figure 2. The conversion sub-system of a mains power supply.**

## WARNING

If any of the circuits shown in this book are to be constructed and investigated, it is the advice of the author and editor that a commercially produced low voltage a.c. supply is used to isolate and protect the experimenter from the domestic 230 V a.c. supply.

Under NO circumstances should an open step-down transformer be used, especially with children in a school laboratory/classroom.

## A.C. TO D.C. CONVERSION — RECTIFICATION

*Half wave rectification.*

The simplest means of converting a.c. to d.c. is by the insertion of a device allowing unidirectional current flow, known as a rectifier, in series with the load (represented by $R_L$).

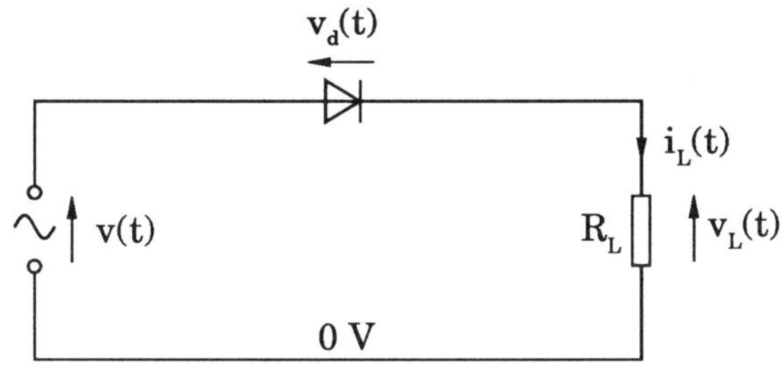


**Figure 3. A half wave rectifier.**

Examination of the resulting current flow in the load (assumed purely resistive) reveals that it is far from d.c. of constant magnitude but consists of a series of pulses, in one direction, known as pulsating d.c. Suppose the a.c. supply is 6 $V_{rms}$ and we consider the lower connection on the circuit diagram to be at 0 V then the upper connection will be oscillating sinusoidally between about –8.5 V and +8.5 V.

$$v(t) = v_p \cos(2\pi f t + \phi)$$

where

$$\begin{aligned} v_p &= 8.5 \text{ V} \\ f &= 50 \text{ Hz} \\ \phi &= \text{arbitrary phase angle} \end{aligned}$$

If the diode is a silicon solid state device it will conduct whenever the potential at its anode is approximately 0.7 V, $V_F$, higher than that at its cathode.

Hence current will flow through the load and there will be a p.d. across it whenever the output from the a.c. supply is greater than +0.7 V or for about half the cycle. The peak p.d. of the pulses, $v_m$, will be about +7.8 V.

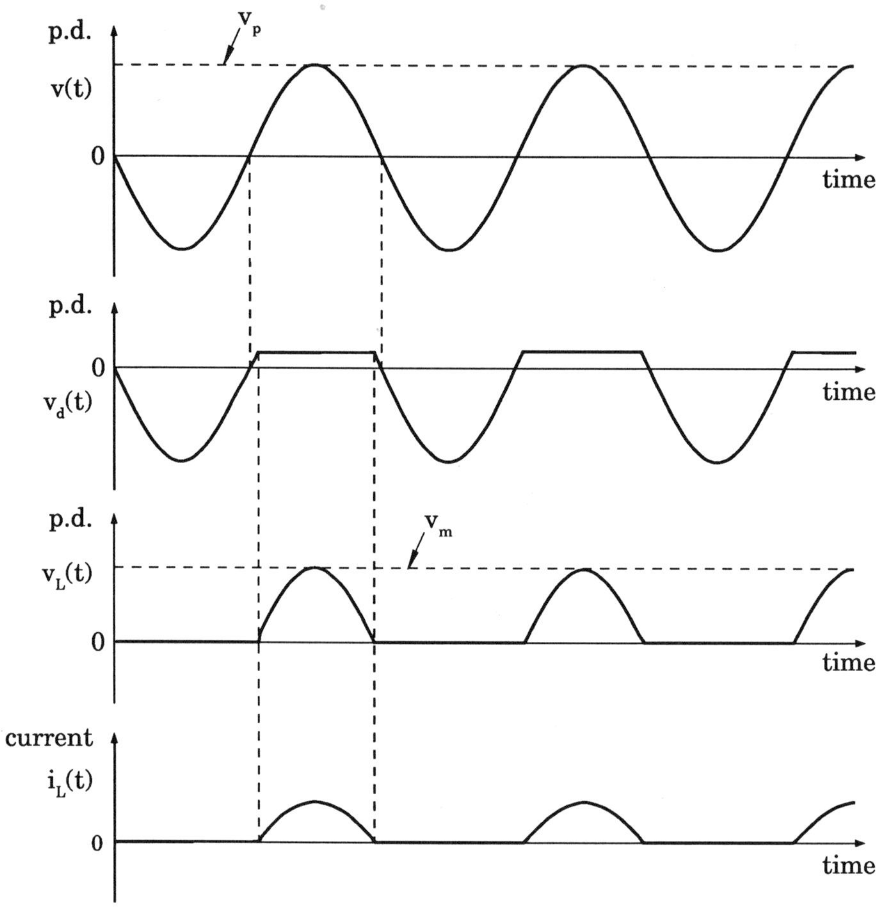


**Figure 4. Voltage and current traces for the ciruit of Figure 3.**

The pulse like nature of the supply can be mitigated by the inclusion of a further sub-system. The common analogy between electric circuits and water supply systems has lead to such a system being referred to as a reservoir system, otherwise, incorrectly, as a smoothing system.

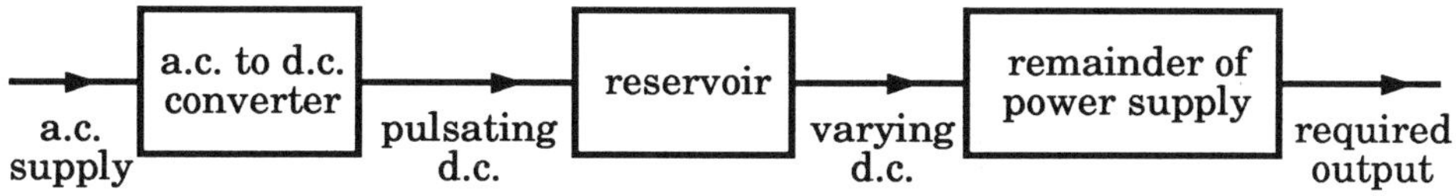


**Figure 5. The reservoir sub-system.**

*Capacitor reservoir circuits.*

The simplest implementation of a reservoir sub-system uses a large capacitor, usually of electrolytic construction, connected in parallel with the load.

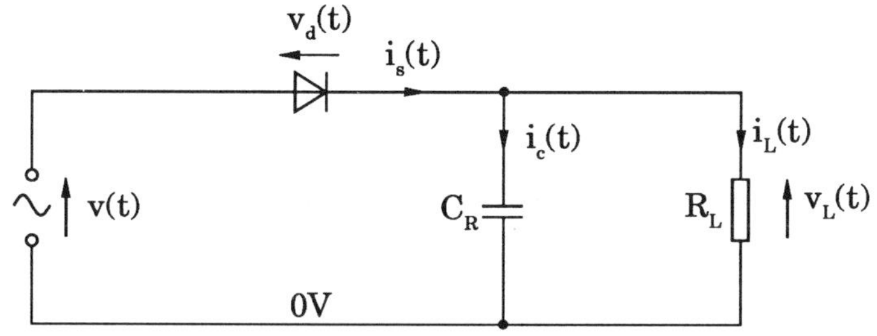


**Figure 6. Half wave rectifier with a reservoir capacitor.**

For some purposes such a supply is adequately "smoothed", for example many early radios and televisions of the so-called "A.C./D.C." variety used this system. The sound output of these devices was frequently accompanied by a 50 Hz hum known as mains hum. This system also sometimes proved disastrous for the local electricity supply as a whole street full of such devices would seriously unbalance the current drawn from the supply and result in breakdown of sub-station transformers.

Analysis of the p.d. present across the load shows that it may considered to be composed of two components, a constant p.d., $v_{dc}$, with an a.c. component, the ripple, $v_r(t)$, superimposed upon it as shown in the sketch graphs shown in Figure 7.

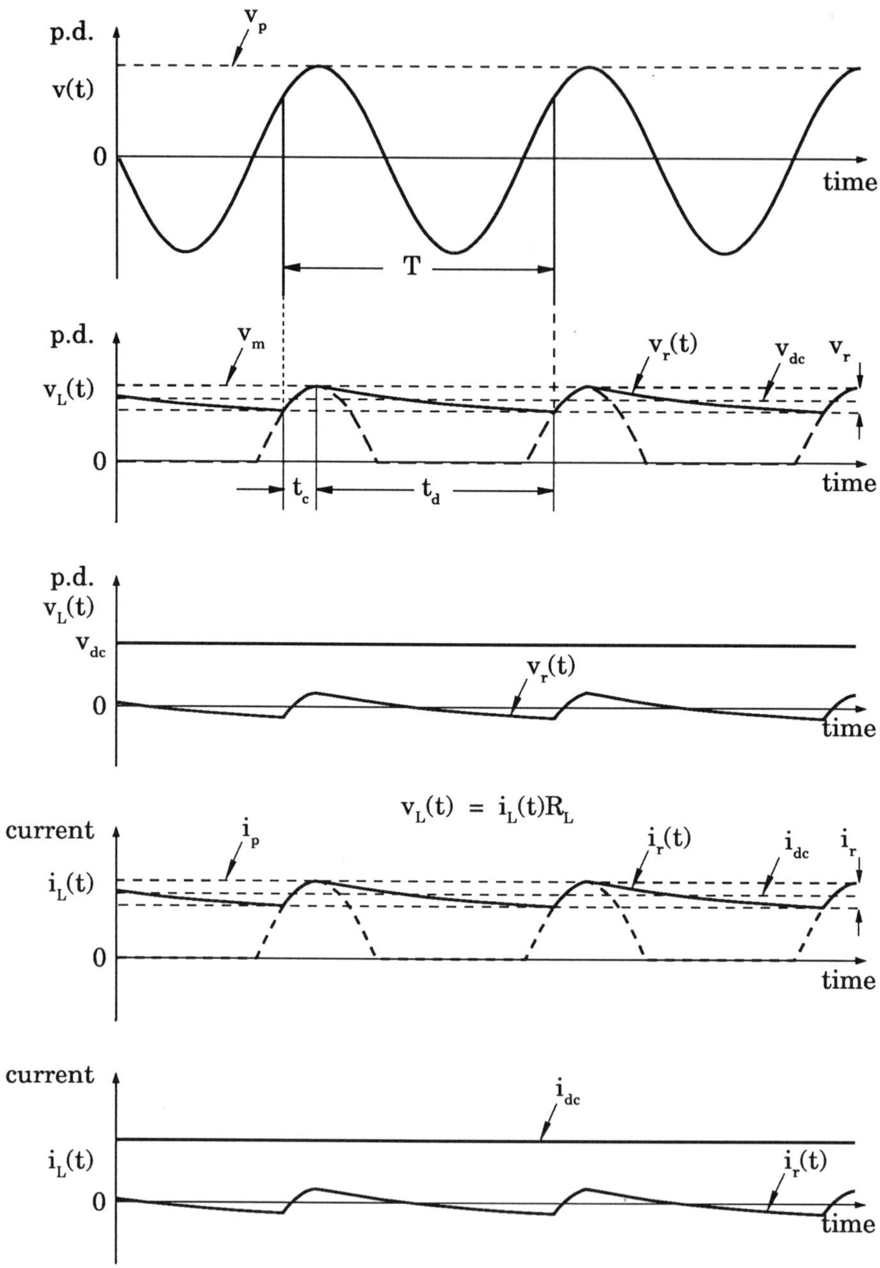


**Figure 7. Graphs of the p.d.'s for the circuit of Figure 6.**

The maximum value of the output p.d. is about 0.7 V less than the peak value of the a.c. input because of the forward voltage drop, $V_F$, in the diode.

It is the ripple p.d., $v_r(t)$, which is responsible for the hum produced by the equipment mentioned above. The ripple can be reduced by "throwing more capacitance at it" but this will be shown to be an inappropriate solution, especially where low ripple content is required.

The capacitor is functioning as a charge storage device. If the time constant $R_L C_R$ is large compared with the period of the supply, T, then the ripple size, $v_{r,}$ will be small. An approximate analysis of the behaviour can now be made if we assume that the time taken to charge the capacitor, $t_{c,}$ is small compared with the discharge time, $t_d$, that $v_L(t)$ is constant, and that all the load current is supplied by the capacitor. If the average current supplied to the load is $i_{d.c.}$ then the charge transferred to the load each cycle is given by

$$Q = i_{dc} \times T$$

Now for a capacitor we have:

$$Q = C \times V$$

where V is the change in potential caused by the charge Q. We may therefore write:

$$C_R \times v_r = i_{dc} \times T$$

or

$$v_r = \left( \frac{i_{dc} \times T}{C_R} \right)$$

Now

$$T = \frac{1}{f}$$

where f is the frequency of the supply,

therefore:

$$v_r = \left( \frac{i_{dc}}{f \times C_R} \right)$$
$$= \left( \frac{v_{dc}}{f \times R_L \times C_R} \right)$$

Suppose we consider a 5 volt , 1 amp power supply, i.e. $R_L = 5\ \Omega$. If we require a ripple voltage of 1% using our 50 Hz supply then:

$$C_R = \left( \frac{\left( \frac{V_{dc}}{V_r} \right)}{f \times R_L} \right)$$

$$= \frac{100}{50 \times 5}$$

$$= 0.4\ F$$

0.4 F is an *extremely* large value of capacitance!

Among the disadvantages of the above circuit are:

(1) we are only drawing energy from the supply during about half the cycle and;

(2) the peak value of the current passing through the rectifier rises rapidly as current is drawn from the power supply.

The peak current is limited only by the impedances of the rectifier and the a.c. supply which in most cases will be a transformer capable of supplying large instantaneous currents. Semiconductor devices capable of carrying the average current supplied by the power supply into the load may not be able to withstand the peak current delivered into the reservoir capacitor during the conduction part of the cycle.

It should be noted that a particularly large current, known as a surge, will flow when the power is switched on and the capacitor is as yet uncharged. As before, this current is determined by the impedance in the mains power supply circuit, but initially there is no p.d. on the capacitor opposing the flow of a current, (i.e. at switch ON the capacitor appears as a short circuit load) and large value capacitors require this current for long periods of time before they charge and offer significant opposition to the current.

The half wave rectification system will not be dealt with in greater detail because of these limitations.

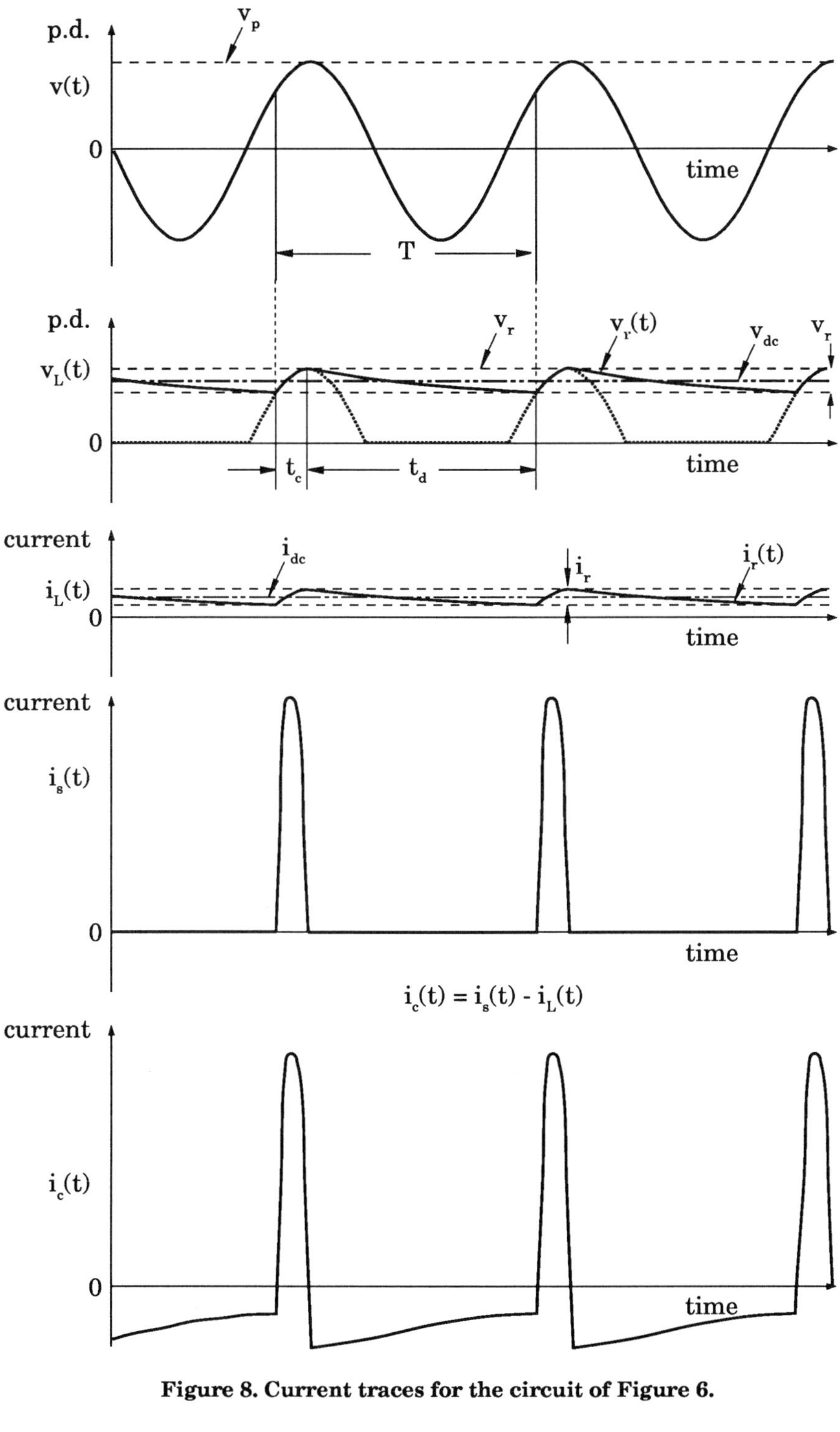


**Figure 8. Current traces for the circuit of Figure 6.**

*Full Wave Rectification.*

A more efficient rectification system than the one shown above uses either a transformer with a centre tapped winding, or four rectifiers in a bridge arrangement to give full wave rectification.

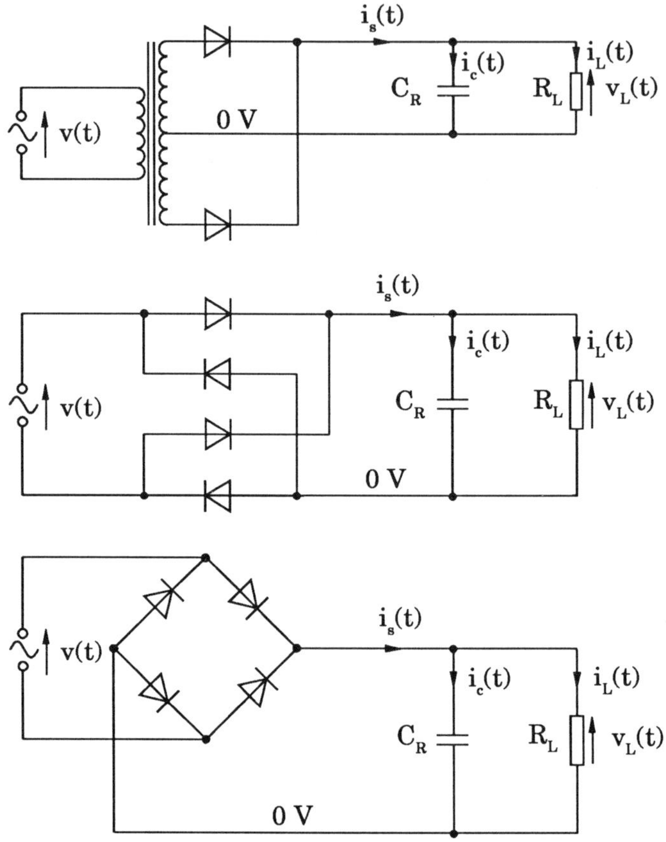


**Figure 9. Full wave rectification circuits.**

The lower two circuits in the previous figure are equivalent, the second being the more traditional representation.

Recent price reductions in the cost of solid state diodes compared with the increasing cost of transformers have resulted in the bridge circuit supplanting the centre tapped transformer circuit.

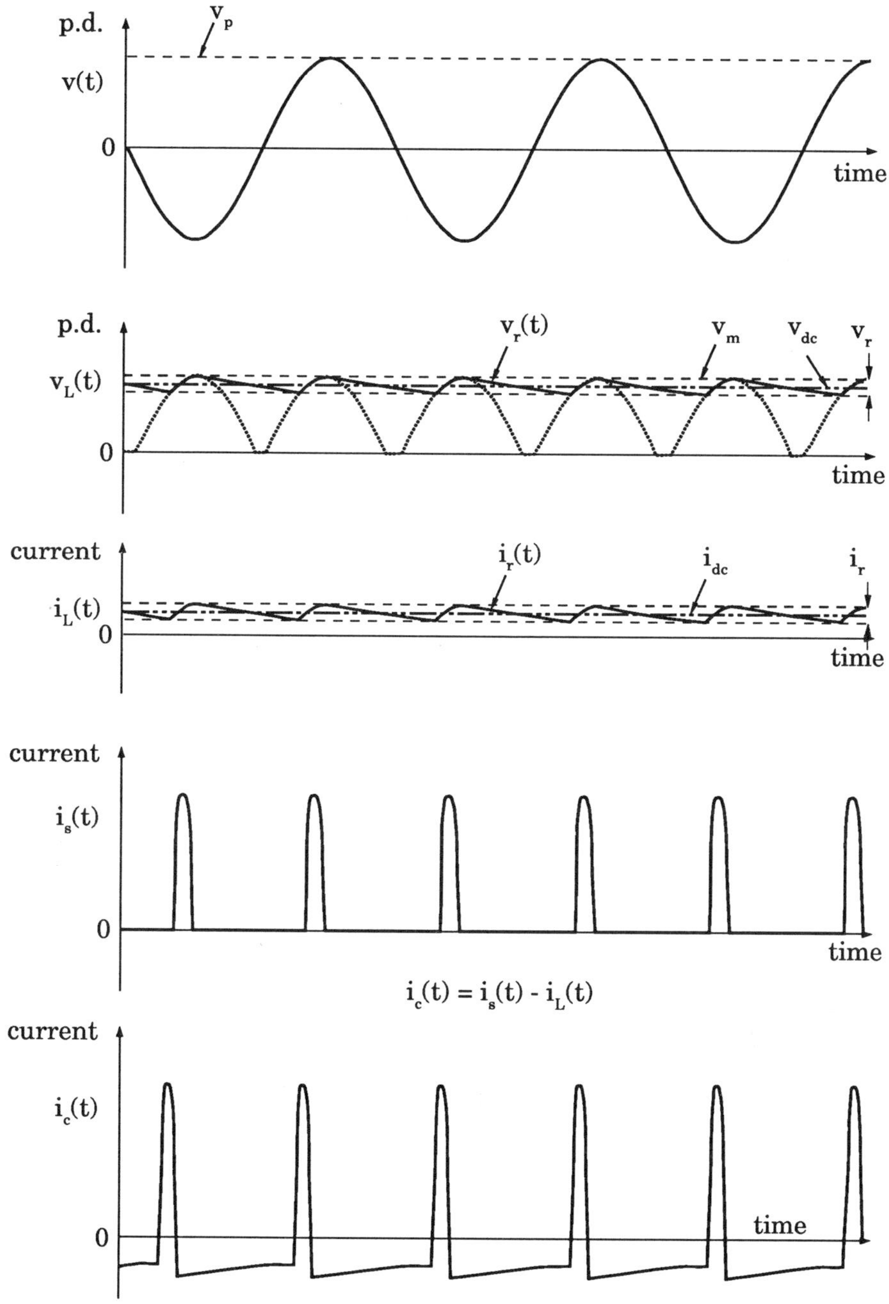


**Figure 10 . Graphs of currents and potentials in the bridge circuit.**

For a transformer with a particular secondary winding which is centre tapped the bridge circuit using the whole secondary will give approximately twice the output p.d. of the circuit using the centre tap and only two rectifiers.

The bridge circuit makes more efficient use of the transformer than the half wave circuit referred to previously as it supplies current for virtually the whole cycle. It should be noted, however that the maximum output p.d. from the bridge circuit is about 0.7 V lower than the maximum output p.d. of a half wave circuit using the same transformer secondary because effectively there are two diodes in series with the load at any time. Graphs of the rectified current, the capacitor current, the output current and the output p.d. in the simple reservoir capacitance smoothing circuit are shown in Figure 10.

Examination of the graphs shows that even using a full wave circuit with a reservoir capacitance a significant ripple p.d. is still present when the power supply is delivering current.

From inspection of Figure 10 we can make the approximation

$$v_{dc} = v_m - \frac{v_r}{2}$$

where $v_m$ is the maximum value of $v_L(t)$ as shown in Figure 10 and $v_r$ is the peak to peak ripple p.d.

If the capacitor discharge time, $t_d$, is considerably less than the time constant, $R_L C_R$ then we may assume that the capacitor discharge current, $i_{d.c.}$, is constant and that the capacitor loses the total charge, $i_{dc}t_d$, which gives a change in p.d.

$$v_r = \left(\frac{i_{dc} \times t_d}{C_R}\right)$$

Again assuming that $t_d$ is much greater than $t_c$ then

$$t_d = \frac{T}{2} = \frac{1}{2 \times f}$$

and

$$v_r = \left(\frac{i_{dc}}{2 \times f \times C_R}\right)$$

$$= \left(\frac{v_{dc}}{2 \times f \times C_R \times R_L}\right)$$

Rearranging, for a particular value of ripple $v_r$, we get

$$C_R = \left(\frac{v_{dc}}{2 \times f \times R_L \times v_r}\right)$$

Substitution of the values used in the previous case gives the value of C as 0.2 F, still an impracticably large value of reservoir capacitor.

*Regulation and supply impedance.*

It has been shown that increasing the current drawn from a simple capacitor filtered power supply results in an increase in the ripple content of the supply. The approximation

$$v_{dc} = v_m - \frac{v_r}{2}$$

was made previously.

Substituting $\left(\frac{i_{dc}}{2 \times f \times C_R}\right)$ for $v_r$ as above we get:

$$v_{dc} = v_m - \left(\frac{i_{dc}}{4 \times f \times C}\right)$$

$$= v_m - (i_{dc} \times r_s)$$

The mean output voltage falls as the load is increased (as the load resistance decreases). It is as if there is a resistance in series with the output. The maximum value of output p.d. $v_L(t)$ when the load is absent (i.e. $R_L = \infty$) is $v_m$ and the effective output (resistance) impedance of the supply, $r_s$, is given by

$$r_s = \left(\frac{1}{4 \times f \times C}\right)$$

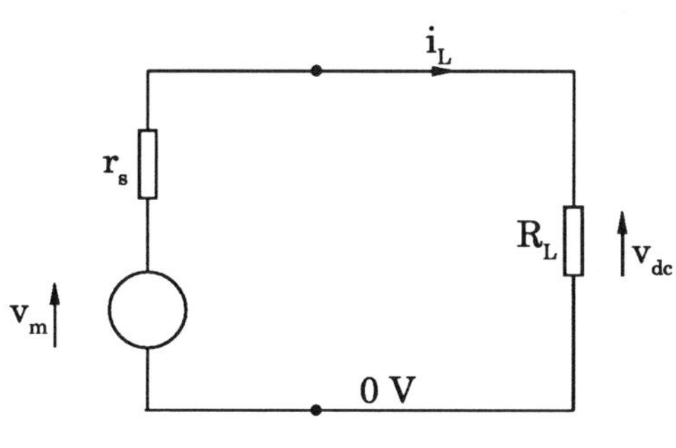


**Figure 11. Impedances in the power supply.**

This is in addition to any impedance introduced by the transformer, the a.c. supply, or the behaviour of the rectifiers. The ability of a power supply to maintain its output voltage while the current taken from it varies is known as its regulation.

By definition:

$$\text{regulation} = \frac{(\text{noload voltage}) - (\text{fullload voltage})}{(\text{fullload voltage})}$$

The regulation is sometimes expressed as a percentage.

It should be noted that any decrease in the ripple content achieved may be at the expense of a reduction in the regulation of the power supply.

*Choke reservoir circuits.*

In the circuits discussed so far the reservoir capacitor is acting as a short term energy store. The energy is stored in the form of an electric field within the capacitor. It is equally possible to use a magnetic field in the same way. The device enabling this to be carried out is the inductor or choke.

The simplest choke reservoir circuit consists of a single inductor in series with the load.

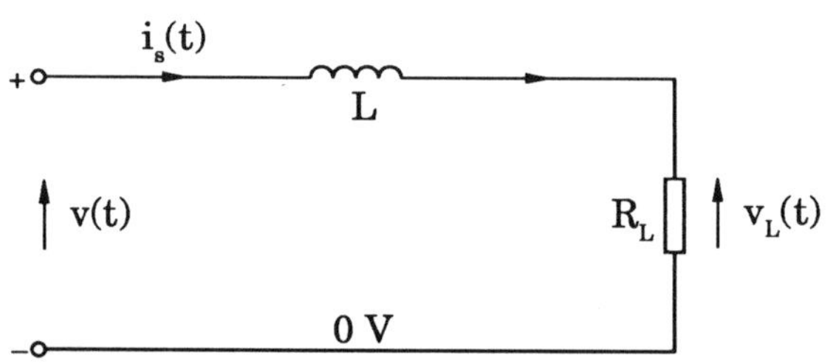


**Figure 12. Circuit with a choke to reduce ripple.**

In theory the use of a suitably large value of inductance will permit the reduction of the ripple current to any required value, however this may not be practical as large real inductances have the disadvantages of expense, weight, bulk, internal resistance and a reduction of their inductance with increasing current.

## FILTERED POWER SUPPLIES

A more feasible approach to ripple reduction involves the introduction of a new sub-system, the low-pass filter.

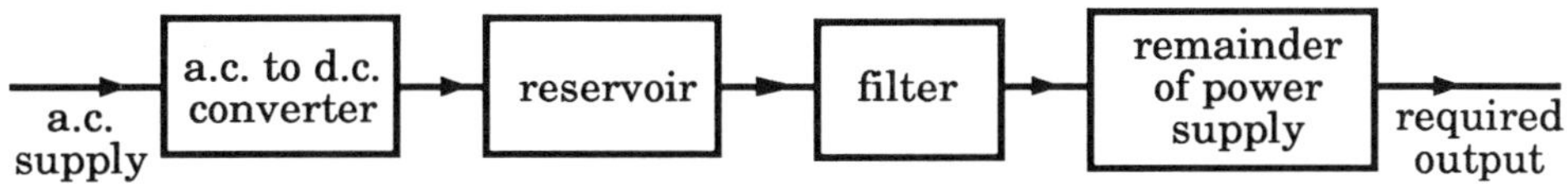


**Figure 13. The position of the filter sub-system.**

One of the easiest ways of producing a low-pass filter is to combine an inductor with a large value capacitor - *a choke input filter.*

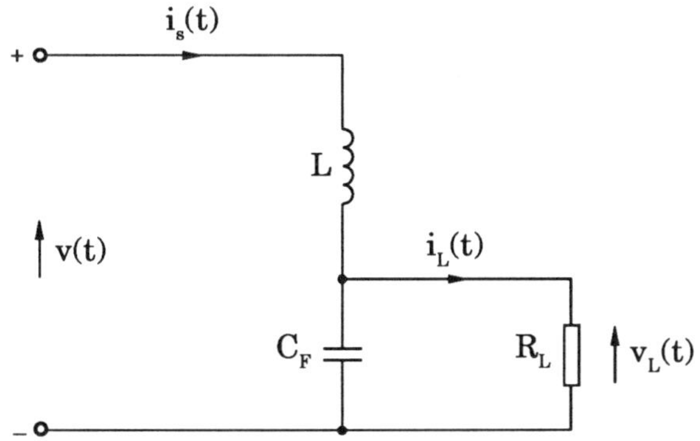


**Figure 14. Circuit with reservoir capacitance and choke input filter.**

It has been shown that when the output p.d. from the full wave rectifier, v(t), followed by the reservoir capacitance, $C_R$, is supplying current to a load, it can be considered as the combination of a constant p.d., $v_{dc}$, and an alternating p.d., $v_r(t)$, having a frequency 2f, twice that of the a.c. supply.

$$v(t) = v_{dc} + v_r(t)$$

Now $v_r(t)$ is not sinusoidal but can be shown to be a combination of a series of sinusoids having progressively increased frequencies and diminished amplitudes. The effectiveness of a low-pass filter rises rapidly with frequency. Therefore if such a filter circuit is shown to be adequately effective at 2f, it will also effectively reduce the higher frequency components present and hence the ripple p.d.

Analysis of the response of the previous circuit is simplified if some assumptions are made:

L is a pure inductance (i.e. it has no resistance);

the reactance of the capacitor $C_F$, $X_c = \left(\frac{1}{2\pi f C_F}\right)$, is small compared to that of the load $R_L$;

the reactance of the inductor L, $X_L = 2\pi fL$, is large compared to the reactance of $C_F$.

Under these assumptions the filter can be treated as a potential divider, i.e. a linear system, so we can apply the principle of superposition and the d.c. and a.c. responses can be considered separately as shown below.

For the d.c. components of v(t), $v_{dc}$:

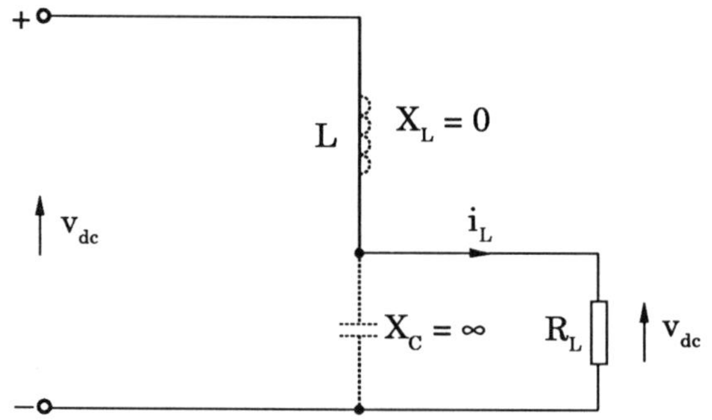


For the (a.c.) ripple component of v(t), $v_r(t)$:

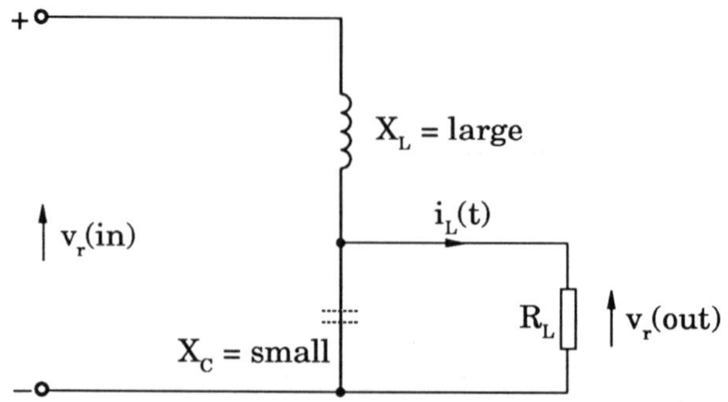


**Figure 15. The d.c. and a.c. circuits of Figure 13.**

The d.c. circuit is quite straightforward. The resistance and reactance of the pure inductance are zero and the reactance of the capacitor is infinite hence there is no attenuation of the d.c. component and the d.c. output of the circuit is $v_{dc}$.

If assumptions 2 and 3 are true then the a.c. circuit may be considered as a potential divider. The reactance of the inductor is $X_L = (2\pi fL)$ and that of the capacitor is $X_C = \left(\frac{1}{2\pi fC}\right)$.

The a.c. output of the potentiometer will then be:

$$v_r(\text{out}) = \left( \frac{\frac{1}{2\pi fC_F}}{\frac{1}{2\pi fC_F} + 2\pi fL} \right) \times v_r(\text{in})$$

$$= \left( \frac{1}{1 + 4\pi^2 f^2 LC_F} \right) \times v_r(\text{in})$$

*Resistor input filters.*

An alternative approach to the low-pass filter sub-system replaces the inductor by a resistor, $R_F$.

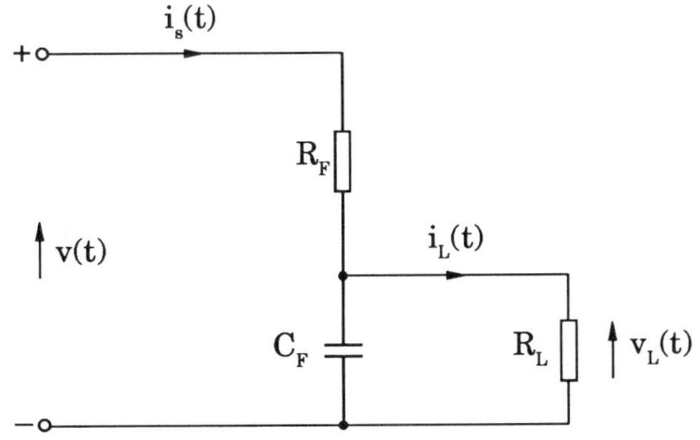


**Figure 16. Circuit with a resistor input filter.**

Resistor input filters were originally rarely used in power supplies as, although they are inexpensive to construct, if a significant reduction in the ripple content is required they result in an appreciable drop in d.c. voltage across the series resistor and the dissipation of large quantities of heat — $R_F$ should be small when compared to $R_L$. However the recent large rise in the cost of inductors coupled with the availability of ever larger value capacitors has led to their increasing use. With the arrangement shown in Figure 16 the ripple is reduced by a factor:

$$\frac{v_r(out)}{v_r(in)} = \left( \frac{1}{\sqrt[+]{1 + (2\pi f R_F C_F)^2}} \right)$$

This resistor input filter may be found in the low voltage supply circuit of many audio power amplifier systems.

# CHOICE OF COMPONENTS

*Rectifier diodes.*

Reference has already been made to the need for the rectifier to be able to safely carry a current far greater than the steady current drawn from the power supply. Usually manufacturers data sheets will give a guide to the application of their products. The device characteristics can be used to assess suitability for a particular circuit. For example the following information is given for the silicon diode type 1N4001:

| Characteristic | Value | Notes |
|---|---|---|
| $V_{RRM}$ | 50 V | Used to be called Peak Inverse Voltage, (PIV), the maximum permissible reverse voltage before failure. This should be greater than $2\sqrt{2}\ V_{RMS}$ with a capacitor input filter, the margin should be at least 50% to ensure reliability. |
| $I_{FSM}$ | 30 A | Maximum instantaneous current e.g. during switch on surge. The value of the switch on surge is very difficult to assess especially for large capacitive filters. Use a generously rated device for reliability. |
| $I_F$(ave) | 1 A | Maximum rectified current. May need to be reduced by 80% for a capacitive load. Once again for reliability allow a 50% margin. |
| $V_F$(max) | 1.1 V @ 1 A | Maximum forward voltage drop when supplying the stated current. |
| $I_R$ | 50 μA @ 50 V | Leakage current when reverse biased by the stated potential difference. |

*Capacitors.*

The large values of capacitance required for reservoir capacitors and low-pass filters in power supplies are usually provided by using electrolytic types. These capacitors are polarised and must not be reverse biased. The characteristics of electrolytic capacitors can vary widely but those of a typical capacitor for power supply use are given below.

| **Characteristic** | **Value** | **Notes** |
|---|---|---|
| Tolerance | –10%,+50% | Note wide variation possible. |
| Temperature range | –55 °C to +85 °C | Limited by electrolyte characteristics. |
| Leakage current | <0.003 μA/μF/V | Depends on capacity and applied voltage. |
| Capacity | 1m5 | 1500μF, but remember the wide tolerance. |
| Voltage | 16 V | This must not be exceeded, it is best to allow a safety margin for reliability. |
| ESR | 107 mΩ | Effective series resistance. The internal (d.c.) resistance of the conducting parts of the capacitor. |
| Impedance | 50 mΩ @ 10 kHz | The a.c. impedance. |
| Ripple current | 2.9 A @ 100 Hz | These are the maximum safe ripple currents. |
| | 5.0 A @ 10 kHz | |
| | both @ 85 °C | Remember, if the ripple current has been removed from the output it must have gone somewhere! This parameter does not usually cause problems. |

*Inductors.*

The relatively high values of inductance required for power supplies lead to the use of either iron or ferrite cored coils. These cores introduce their own characteristics. A result is a decrease in the inductance with rising current through the coil. For example a choke having and inductance of 25 H at 50 mA has an inductance of 5 H at 250 mA. Large values of inductance tend to be both expensive and heavy especially where current flow is measured in amps rather than milliamps. As long as the inductor is rated for the required current output there should be no problems.

*Resistors.*

The only difficulties usually encountered with resistors in power supplies are caused by overheating due to the use of components with an inadequate power rating. It is wise to allow a generous power supply rating remembering that adequate ventilation will also be required.

High wattage resistors should **not** be mounted close to printed circuit boards, but instead arranged in an adequate air flow: if necessary an *appropriate* heat sink should be used.

## REGULATED POWER SUPPLIES

The power supply circuits examined so far may be designed to provide any required voltage and current with a stated ripple content, but they may turn out to be physically large, heavy and expensive to construct. None of these circuits can provide regulation in the face of fluctuations of the a.c. supply.

Where the required power supply is defined within close tolerances it becomes more practical to adopt a simple capacitance reservoir system to give what one might term raw d.c., having poor regulation and high ripple content, and then control the output with electronic circuitry using active devices. This is known as a regulated or voltage stabilised power supply. The function of a regulator circuit may also be to limit the current drawn to safe levels and automatically shut down the power output in the event of equipment temperatures rising too high. Standard TTL logic gates are the devices needing the most closely regulated power supplies commonly found in schools. They require between 4.75 V and 5.25 V, hence the suitability of a circuit for supplying standard TTL devices will be made be reference to.

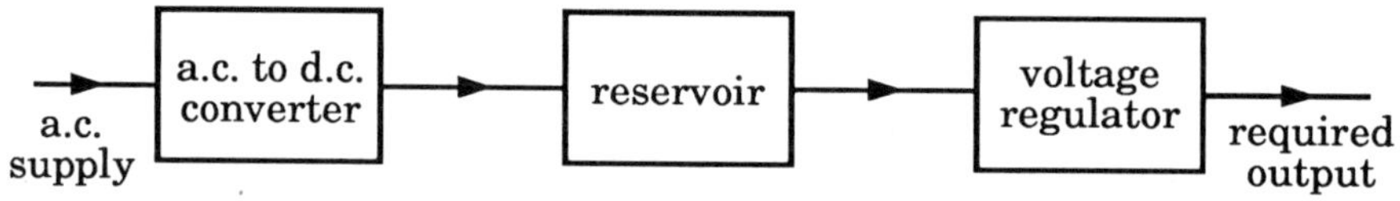


**Figure 17. The power supply system with a regulator sub-system.**

*Zener diode regulated power supplies.*

When a semiconductor diode is reverse biased with an increasing potential difference it will eventually breakdown and conduct. Diodes which are designed to operate non-destructively in the reverse breakdown region are called zener diodes.

*Shunt regulated circuits.*

The simplest form of regulator, for use where moderate currents are required and the specifications are not too tight, utilises the zener diode. A suitable circuit is shown in Figure 18.

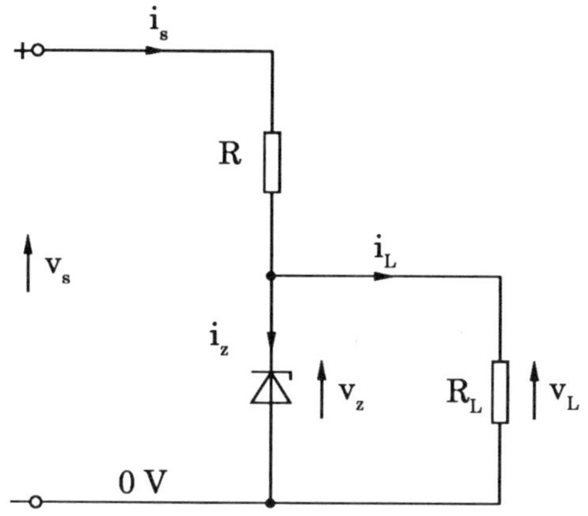


**Figure 18. The basic zener diode circuit arrangement.**

The operation of the zener diode circuit may be analysed more easily by a consideration of the p.d./current relationship of the zener diode. A typical zener diode characteristic curve is shown below.

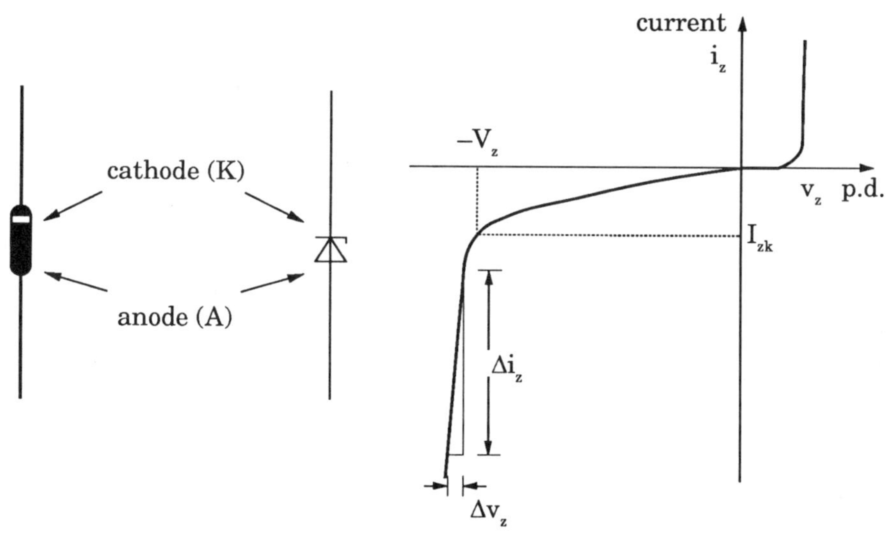


**Figure 19. The zener diode, circuit symbol and its characteristic curve.**

The zener diode behaves like a normal diode when biased in the forward direction but becomes conductive when the reverse bias p.d. exceeds the value $V_z$ known as the Zener voltage which can be between 2.4 V and 75 V, according to the diode.

Ideally (for perfect voltage regulation) the slope of the breakdown region of the diode would be vertical but in practice the slope is oblique, as shown, and is described by

$$\Delta v_z = R_z \times \Delta i_z$$

where $R_z$ is the *dynamic resistance* (or *slope resistance*) of the device. The lower the value of this resistance the better the regulation provided by a power supply using it.

The operation of the above circuit, assuming that $R_z$ is zero and that $v_s$ is greater than $V_z$ may be described by the equation

$$i_z = i_s - i_L$$

$$= \left(\frac{v_s - V_z}{R}\right) - \left(\frac{V_z}{R_L}\right)$$

As the load current increases the zener current, $i_z$, decreases. The minimum value of zener current at which regulation is maintained is $I_{zk}$. See Figure 19. The current output range from a particular circuit will be limited by the maximum zener current permissible, this is in turn determined by the maximum power dissipation of the type of zener employed. Typical power dissipations for zener diodes are 0.5 W, 1.3 W, 5 W and 20 W. Zener voltages range from 2.4 V to about 75 V.

If a non-standard value of zener is required zener diodes may be connected in series, in which configuration their zener p.d.s are added together.

Let us now consider the component values required to construct a regulator, using the circuit of Figure 18 with a rectifier and capacitor filter (as in Figure 9), to give a regulated supply of 5.1 V at up to 100 mA shown in Figure 20.

The maximum output voltage from the smoothing capacitor is

$$v_m = v_p - 2 \times V_F$$

$$= 7.8 \text{ V}$$

Where $v_m$ is now the maximum value of $v_s(t)$, v(t) is the a.c. supply of peak value, $v_p$, and $V_F$ is the forward voltage drop for the rectification diodes.

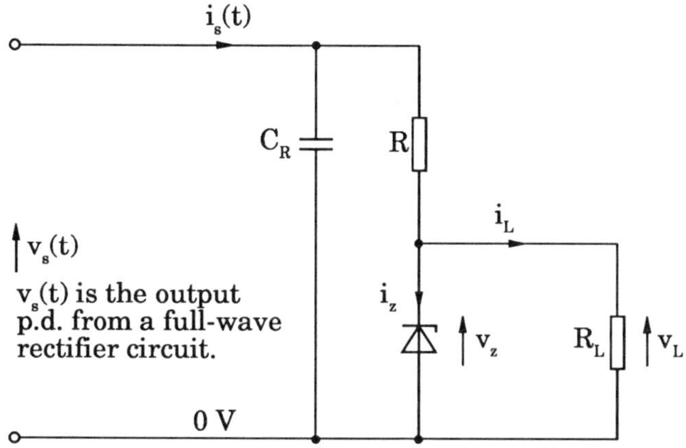


**Figure 20. A zener stabilised 5.1 V power supply.**

Suppose we make a wild guess that the 1m5 (1500 μF) capacitor discussed above will be suitable for the smoothing capacitor and that including the minimum zener current provision the current drawn from the rectifiers is 110 mA. From earlier, (page 12) for a full wave rectification circuit followed by a reservoir capacitor, it was shown that:

$$v_r = \left( \frac{i_{dc}}{2 \times f \times C_R} \right)$$

where $i_{dc}$ is the average value of $i_s(t)$ (= $i_{dc} + i_r(t)$).

In this case

$$v_r = \left( \frac{0.11}{2 \times 100 \times 0.0015} \right)$$

$$= 0.37 \text{ V}$$

and

$$v_{dc} = v_m - \frac{v_r}{2}$$

$$= 7.62 \text{ V}$$

Further we notice that $v_s(t)$ must not fall below the zener voltage, $V_z$.

$$\left( v_{dc} - \frac{v_r}{2} \right) > V_z$$

This is essential if the zener is to remove as much ripple as possible.

R can now be calculated:

$$
\begin{aligned}
R &= \frac{V}{I} \\
&= \left(\frac{7.62 - 5.1}{0.11}\right) \\
&= 23\ \Omega
\end{aligned}
$$

Only preferred values of resistor are usually available, but the 22 Ω value will be well within our tolerance range.

The power dissipated in R is $i_R^2R$ or 0.26 W. A 0.5 W rating for R is suitable.

The power dissipated in the zener has a maximum value when the power supply output current is zero, ($v_z i_z = 5.1 \times 0.11 = 0.56$ W). So a 1.3 W device would give a good safety margin. A suitable choice would be the BZX85C5V1 which has a dynamic resistance, $R_z$, of 10 Ω . Reference to the appropriate data sheet reveals that the zener voltage is tested at 45 mA zener current.

In our example, as the current drawn from the power supply increases to 100 mA the zener current falls to 10 mA so the output voltage falls by 0.035 x 10 = 0.35 V, maximum, to 4.75 V. It can also be seen that when the power supply is supplying minimum current the output voltage will rise to 5.75 V.

Regulation has already been defined. We have:

$$
\begin{aligned}
\text{regulation} &= \frac{(\text{no load voltage}) - (\text{full load voltage})}{(\text{full load voltage})} \\
&= \frac{1}{5.1} \\
&= 20\%
\end{aligned}
$$

The ripple p.d. $v_r$, will cause a ripple current of amplitude $i_r$ through the zener is given by

$$
\begin{aligned}
i_r &= \left(\frac{v_r}{R + R_z}\right) \\
&= \frac{0.37}{33} \\
&= 0.011\ \text{A}
\end{aligned}
$$

This will give rise to a ripple on the output of $i_r R_z = 0.11$ V

Hence the ripple content of the output is now about 2%. The variation of supply voltage calculated exceeds that recommended for safe operation of TTL circuits. If TTL loads are to be supplied a zener with a smaller slope resistance, such as a 5 W type might be employed, but the different working point required will necessitate a recalculation of circuit component values.

In our earlier discussions we assumed that all the current supplied by a reservoir circuit came from the capacitor. Even if this is the case, here, the ripple current will be well below the maximum value given for the capacitor so it should operate safely.

The zener diode regulator is operating in a parallel with the load, sometimes referred to as a "shunt", circuit. Any current not drawn by the load passes through the zener, hence the output current cannot be regulated over a greater range than the maximum current carrying capability of the zener diode. The results calculated above represent in some ways a worst case for the implementation of a zener diode regulated supply. Where the maximum current demand is considerably less than the 100 mA budgeted for in the example both regulation and ripple factors would be considerably improved. However, where the range of current demand is wider the situation rapidly worsens and some form of series control of the output current becomes necessary.

*Series Regulator circuits.*

The simplest series circuit utilises a transistor or darlington transistor capable of carrying the maximum current required by the load.

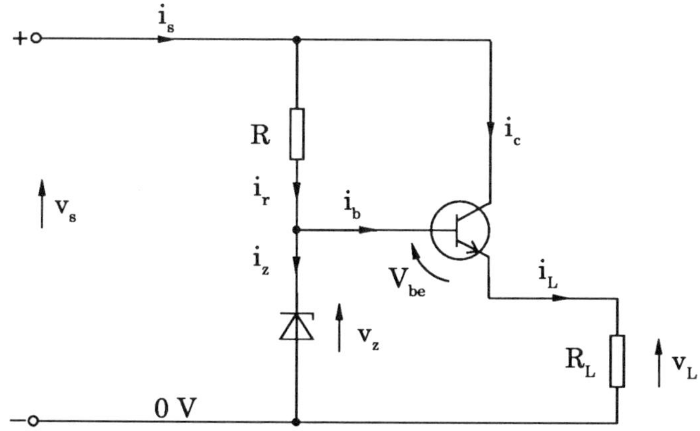


**Figure 21. Simple zener stabilised series transistor regulator circuit.**

Take $v_s$ as the average value of the supply input p.d., (= $v_{dc}$, $v_r(t)$ assumed to be small) and $i_s$ as the average value of the supply current, (= $i_{dc}$, $i_r(t)$ assumed to be small).

The transistor in this circuit is operating as an emitter follower, i.e. the emitter (output) potential, $v_L$, is determined by the potential at the base, $V_z$, determined by the zener diode. We note, though, that the input characteristics of a typical silicon bipolar transistor show that the transistor conducts when $V_{be}$ is around 0.7 V (or 1.4 V for the darlington). Hence $v_L$ will be about 0.7 V less than $v_z$. In our example we required a 5.1 V supply. Using a 5.1 V zener will result in an output of about 4.4 V. A preferred zener value of 5.6 V is available, giving 4.9 V, possibly a bit low for TTL devices allowing a margin of 0.15 V above the minimum 4.75 V as current is drawn by the load . The next preferred value is 6.2 V, giving 5.5 V: dangerously high for use with TTL loads. It is possible to use two zeners, a 2.7 V and a 3.0 V in series. Another solution is to use the forward voltage drop of a silicon diode which should match the drop across the transistor junction.

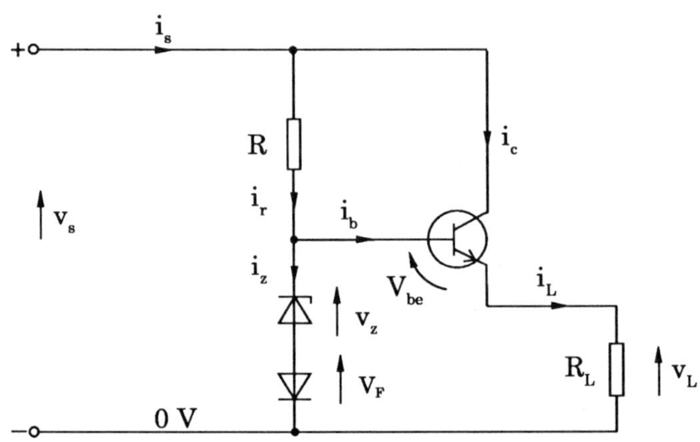


**Figure 22. Circuit showing use of silicon diode to correct for voltage loss in the series transistor.**

The base current, $i_b$, drawn by the transistor will depend on the current drawn by the load, $i_L$, and the gain of the transistor. As before the resistor R is selected to supply sufficient current to keep the zener operating on the linear part of its characteristic whatever base current flows. When selecting a transistor allowance must be made for variations in the gain between transistors of the same type. The transistor must be able to dissipate the heat generated when the load draws its maximum current.

Let us consider the supply of 5.1 V at 100 mA again, using the circuit of Figure 22 with a 5.1 V zener.

The choice of transistor may seem fairly arbitrary, as many types will be satisfactory. A common strategy is to first list devices capable of passing the required current and dissipating sufficient power and then to select for adequate gain at reasonable cost. The choice should not be limited to straightforward bipolar types, but Darlington transistors and Field Effect Transistors (FETs) should also be considered.

A useful general purpose bipolar device is the 2N3053, a silicon NPN low cost transistor with the following characteristics:

| | | |
|---|---|---|
| $P_T$ | 800 mW | Maximum power dissipation. |
| $I_c$ | 1 A | Maximum collector current. |
| $V_{CEO}$ | 40 V | Maximum collector-emitter voltage. |
| | | ($I_c \times V_{CEO}$ must be less than 0.8 ($P_T$)). |
| $V_{CBO}$ | 60 V | Maximum collector-base voltage. |
| $h_{FE}$ | 50-250 | Current gain range (common emitter). |
| $f_T$ | 100 Mhz | Maximum useful working frequency. |

The lowest current gain expected for the device is ×50 hence the base current, $i_B$, will rise from 0 to a maximum of 2 mA as the current drawn by the load, $i_L$, rises from 0 to 100 mA.

We have

$$v_s(t) = v_{dc} + v_r(t)$$

$$i_s(t) = i_{dc} + i_r(t)$$

where $v_r(t)$ and $i_r(t)$ are the supply ripple voltage and current respectively, as previously.

The zener current could be set to about 10 mA to ensure that it is always on the linear part of its characteristic.

The required value of R is

$$R = \frac{v_{dc} - (v_z + V_F)}{i_R}$$

$$= \left(\frac{7.62 - 5.8}{0.01}\right)$$

$$= 182\ \Omega$$

A 180 Ω 0.25 W resistor will do nicely.

Note: $i_z$ will vary, from 10 mA ( $i_L = i_b = 0$), to 8 mA ($i_L$ = 100 mA, $i_b \leq 2$ mA).

Power dissipation in the zener diode is now only around 50 mW so the smallest device available will be suitable. The maximum power dissipation in the transistor occurs at maximum current output when:

$$\begin{aligned} P &= i_L \times (v_{dc} - v_L) \\ &= 0.1 \times (7.62 - 5.1) \\ &= 0.25\ W \end{aligned}$$

which is well within the transistor's capabilities. A suitable silicon diode would be type 1N4148 which is a low cost switching device.

The variation in $v_z$ as the load current, $i_z$, varies from minimum to maximum is now

$$\begin{aligned} \Delta v_z &= \Delta i_z \times R_z \\ &= 0.002 \times 10 \\ &= 0.02\ V \end{aligned}$$

so we now expect the regulation of the output to be better than 1% and the ripple content to be smaller. Unfortunately we have not considered the variation of $V_{CE}$ across the transistor with current through it or with the change in temperature of the device or variation in $V_F$ with $i_z$. Measurement of the regulation obtained in practice with such a circuit makes an instructive exercise.

Further improvement in the regulation of the output obtained requires sensing of changes in the output and their comparison with a reference voltage so that compensating adjustments can be made. In other words we need to use a negative feedback control system.

The necessary negative feedback control system can be simply implemented using a high gain differential amplifier.

*A regulated power supply employing series negative voltage feedback.*

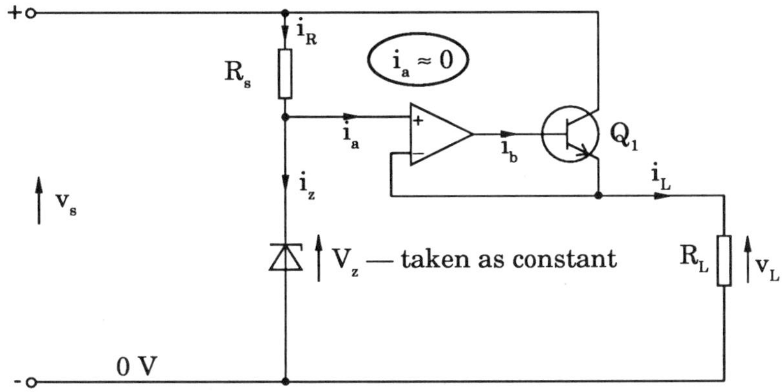


**Figure 23. Regulated power supply with feedback.**

In the circuit above any incipient change in the output voltage, $v_L$, perhaps caused by a change in the load current, $i_L$, is compared with the reference voltage $V_z$. The high gain of the differential amplifier ensures that an adjustment to the base current of the series transistor $i_b$ is made to virtually eliminate the change in $v_L$. (A discussion of the operation of negative feedback with high gain amplifiers will be found in the text "A Look at The Operational Amplifier" by David S. C. Thompson which is available from the publisher of this text). Further, the high input impedance of the amplifier results in the zener current being varied only by changes in $v_s$ and its associated ripple content, if any.

If the proportion of the output voltage fed back to the op-amp is varied then the output voltage can be varied.

Provided that the gain of the differential amplifier is sufficient, then regulation of the output voltage will be maintained between the minimum voltage output, which is determined by $V_z$ when $v_L = V_z$, and the maximum output when $v_L = (v_s - V_{CE})$ for the series transistor. Note that the ripple content of the output will also be minimal as long as

$$v_L < v_s - \left(V_{CE} + \frac{v_r}{2}\right)$$

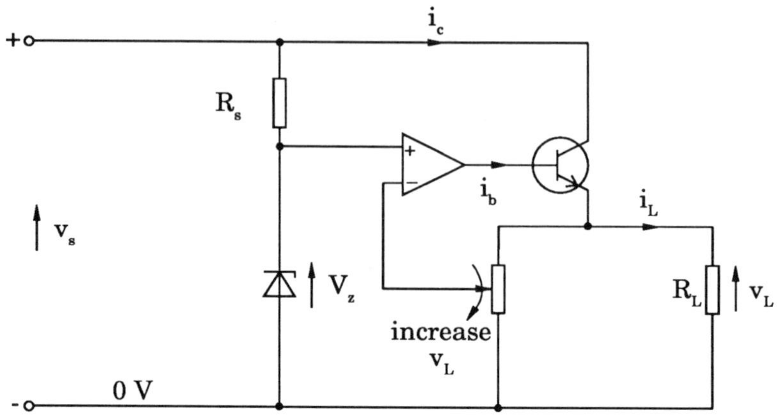


**Figure 24. Variable regulated power supply.**

The regulation of this power supply is sufficiently good for the variation of the zener voltage, $V_z$, with temperature to be noticeable, for zener voltages around 5 V the temperature co-efficient of $V_z$ is approximately +1 mV/°C depending on the type.

The circuits above could be used for our example of 5.1 V 100 mA supply utilizing a 2N3053 transistor and a CMOS ( 7621 ) op-amp. (The μA741 op-amp is **not** suitable as the output can get to only within about 2 V of the supply rail, so the 7 V (approx) of $\left(v_s - \frac{v_r}{2}\right)$.

will not allow the output to rise to the 5.8 V required at the transistor base.) The circuit of Figure 24 could be used with a zener voltage of less than 5.1 V allowing $v_L$ to be trimmed to exactly the required value using the potentiometer.

*Current overload protection.*

None of the power supplies discussed so far incorporate protection against excessive current demand by the load. This can only be implemented by providing a current sensing sub-system which is able to control the output current and over-ride the voltage control sub-system already discussed.

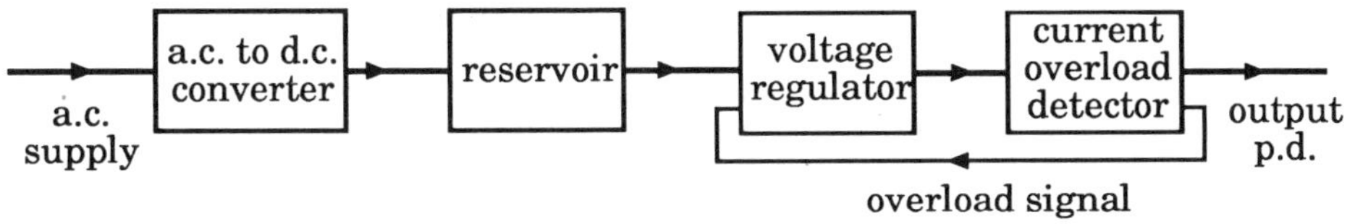


**Figure 25. A power supply system with overload protection.**

A circuit with current overload protection is shown Figure 26.

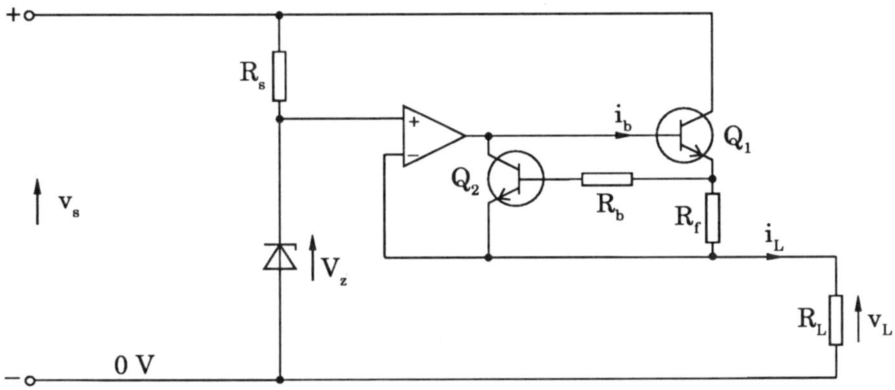


**Figure 26. A current overload protected power supply circuit.**

The current overload protection function is provided by the transistor $Q_2$. The output current, $i_L$, now passes through a low value resistor $R_F$. As $i_L$ increases the p.d. across $R_F$ rises until it reaches about 0.7 V at which point transistor $Q_2$ starts to conduct diverting any increase in the base current of the series transistor $Q_1$, $i_b$. The resistor $R_b$ provides current protection for $Q_2$. The magnitude of the maximum output current is fixed by the value of $R_F$.

It is quite feasible to add circuitry to limit the output current in a similar way in the event of possibly deleterious rises in temperature. Such a project would be of only academic interest as such circuits are cheaply available fabricated on single silicon "chips" as monolithic regulator integrated circuits.

*Monolithic regulator devices.*

A glance at any electronics supplier's catalogue will reveal a ready made regulator device for almost any conceivable function. The application of some of the less expensive and most readily available devices will be considered here. The simplification achieved by the use of these i.c.s can be gauged from the system diagram showing their application.

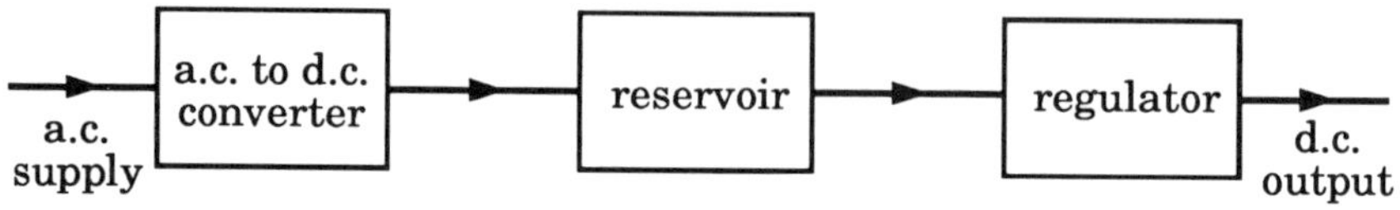


**Figure 27. The power supply system using a monolithic regulator.**

Regulator devices are available in a number of series, each member of the series having the same function and power rating but giving different output voltages. The most commonly met with series is the 78 series of voltage regulators giving up to 1 A at between positive 5 V and 24 V, the 7805 and 7824 respectively. A complementary range of devices, the 79 series gives a negative output. For low current applications the 78L and 79L series take up less space and offer a cost advantage. For high current applications the 78S and 79S series are available. All these devices are fully overload, short circuit and thermally protected. The only circumstances they are not protected against are the application of inputs with excessive voltage or incorrect polarity and over voltage input to the output terminals.

The characteristics of all monolithic regulators devices are broadly similar. As an example let us once again consider a 5 V supply, but this time at up to 1 A output current. A suitable device would be the 7805. The characteristics of the 7805 are given overleaf.

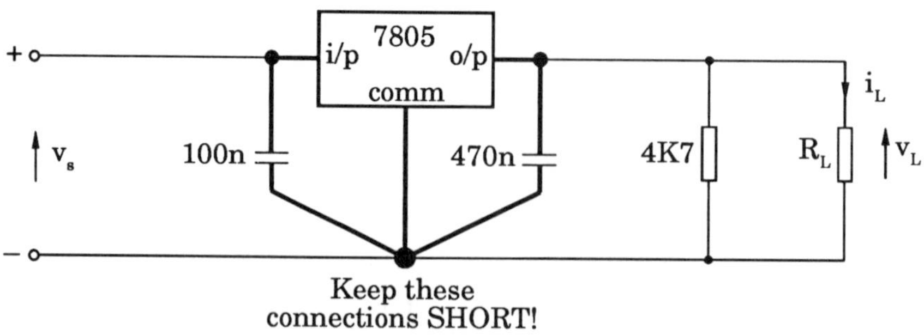


**Figure 28. A fixed output voltage regulated power supply circuit using a monolithic integrated circuit.**

| Characteristic | Value | Notes |
|---|---|---|
| Input voltage range | 7-25 V | The input must be at least 2 V above the output to maintain regulation. This includes variations like increased ripple under load. |
| Load regulation | 0.2% | |
| Ripple rejection | 70 dB | The ripple voltage is reduced by a factor of about 3000. |
| Output resistance | 30 mΩ | |
| Output noise voltage | 0.04 mV | Between 10 Hz and 100 kHz. |
| Short circuit current | 750 mA | |
| Max junction temp | 125 °C | For maximum output a heat sink will be required. Preferably better than 4 °C/W. |

A typical circuit application is shown in Figure 28. The capacitors and resistor associated with the integrated circuit in the above circuit are components which cannot be satisfactorily incorporated on the chip. The circuit reproduces the functions of that in Figure 26. with the addition of thermal protection. In our original specification we required an output at 5.1 V. Although the circuit shown in Figure 28 would normally suffice, it is possible, if required, to increase the output voltage but with the drawback of steadily decreasing regulation.

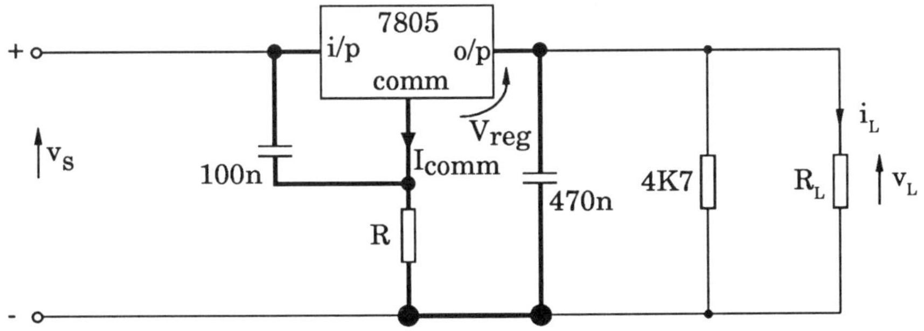


**Figure 29. Circuit to give increased output voltage.**

For this circuit we have $v_L = V_{reg} + I_{comm}R$ where $V_{reg}$ is the regulator voltage and $I_{comm}$ is the regulator quiescent current, **both** assumed to be constant. In the case of the 78 series $I_{comm}$ is about 1.5 mA.

It is possible to use the same chip as a constant current source, for example when charging nickel cadmium rechargeable cells, or as a constant current source in sensor circuitry. The regulator maintains a constant p.d. across the resistor, R, which is in series with the load, and hence maintains a constant current in the load.

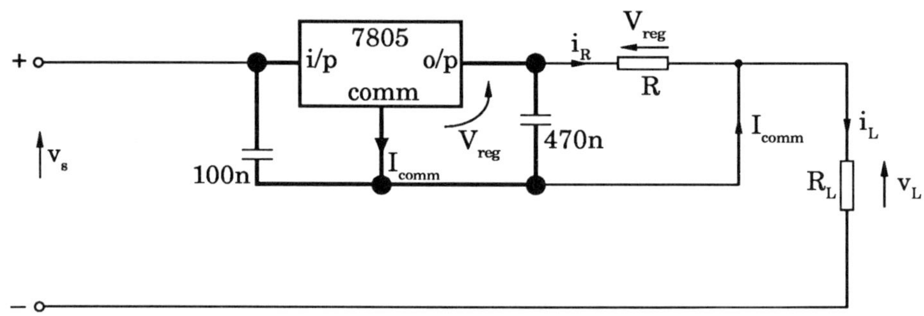


**Figure 30. Circuit of a constant current generator.**

Within the constraints of the possible output voltage range and the maximum power dissipation of the device, the current output $i_L$ is given by

$$i_L = \left( \frac{V_{reg}}{R} + I_{comm} \right)$$

*Switching regulators.*

The series regulator systems discussed so far are inherently inefficient, there is a potential drop across the regulator and as current is drawn by the load, energy must be dissipated. For example, the maximum possible efficiency of a TTL system regulator operating from a raw 10 V supply is 50%. The waste energy is dissipated as heat, which can lead to cooling problems when current demand exceeds about 1 A. The switching regulator, the basis of the so-called "switch mode power supply", overcomes this problem by transferring charge from the raw supply as a regular series of current pulses which are integrated to form the output. The supply is regulated by controlling the duration of the current pulses, as more current is drawn from the output of the supply system the duration of each pulse is increased. A system employing this method is shown in Figure 31.

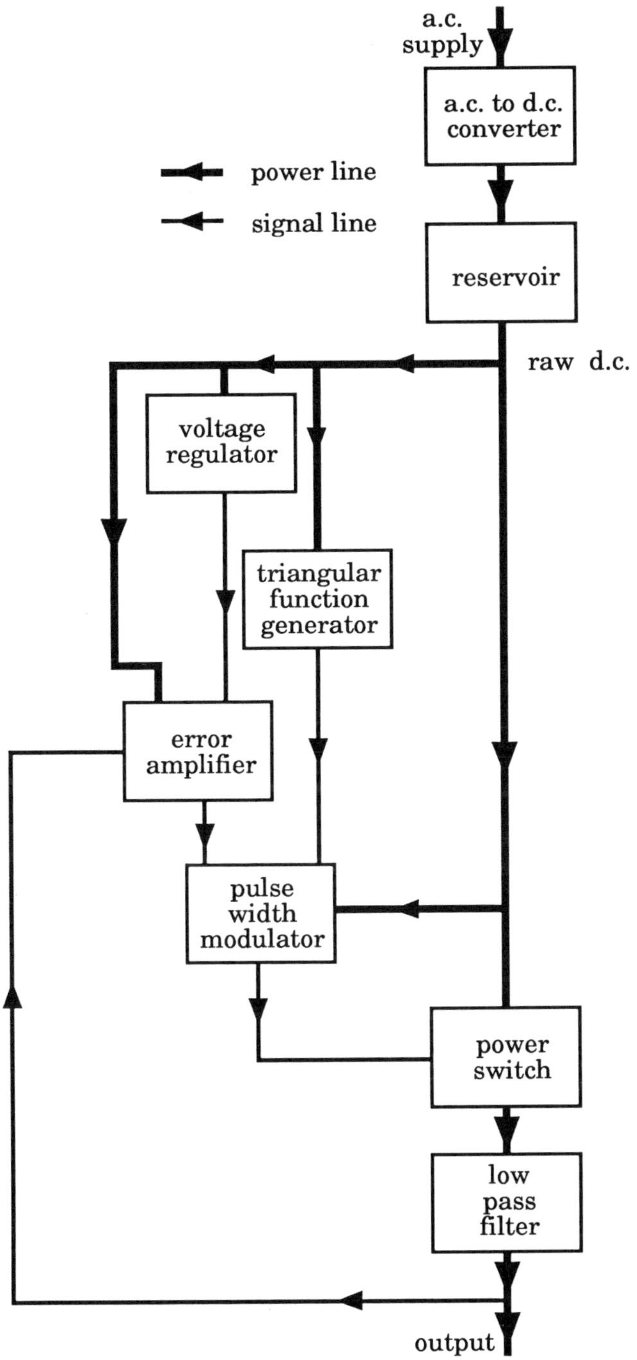


**Figure 31. A switching regulator system.**

In the switching regulator system the output p.d. is continuously substracted from the p.d. of a reference regulator by a differential amplifier to produce an amplified error signal which is considerably greater in amplitude than the actual error. This is shown as the error amplifier in the system diagram. The output p.d., of period T, from a triangular function generator supplies a signal to the non-inverting input of a comparator which has the output of the error amplifier applied to its inverting input. The comparator acts as a pulse-width modulator varying the duration of the pulses in response to the error signal. An error signal indicating too great an output p.d. results in pulses of shorter duration, as shown in Figure 32.

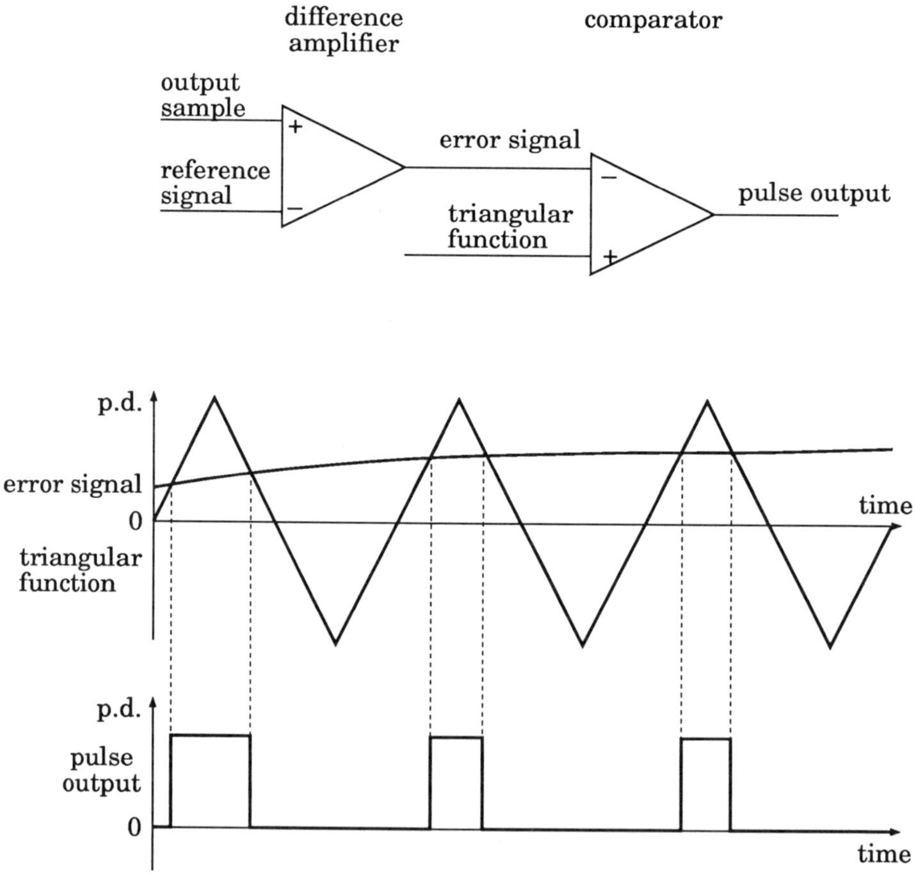


**Figure 32. A pulse width modulator and its action.**

The output pulses are applied to a power switch, perhaps a power bipolar transistor or a power MOSFET, controlling the flow of charge from the raw supply to the output low pass filter which acts like an *electrical flywheel* with the capacitance as the *mass*.

The output filter consists of a choke input filter as analysed previously (see page 15). If the reactance of the capacitor is much smaller that that of the inductance at the fundamental frequency of the function generator then all harmonic components of the output of the power switch, which can be considered as a "square wave" will also be considerably attenuated and the output p.d. will be the average value of the square wave. The values of inductance and capacitance required for the output filter may be significantly reduced by the choice of a sufficiently small value of period T for the output p.d. from the function generator, i.e. $\sqrt[+]{LC} \gg \left(\frac{T}{2\pi}\right)$.

The limiting factor for the reduction of the period T will probably be the switching speed of the power switch device.

*Regulation of battery power supplies.*

Although apparently offering the advantage of simplicity the use of batteries may not be satisfactory without the provision of control circuitry.

Most of the difficulties occur as batteries age and their internal resistance rises. As charge is taken from a battery there may also be a drop in terminal p.d. If the change in terminal potential is not important the effects of the increase in internal resistance may be reduced by using a large capacitance in parallel with the battery. This may be particularly effective in overcoming instability or distortion in audio amplifiers.

Where the drop in supply voltage is critical, one of the voltage regulator circuits discussed above must be employed. It should be noted that this will considerably increase the voltage of the battery required as the voltage at the required discharge end-point must exceed the output of the regulator circuitry by, typically, 2 V. All regulator circuits increase the current drawn from the raw supply. Where minimum power consumption is required the regulator type 7663 may prove of use, being of CMOS construction it requires less than 4 μA operating current, but its maximum output current is limited to 25 mA.

# VOLTAGE CONVERSION

The production of a single stable source of direct current from a raw rectified d.c. supply has been discussed. Where two or more supply potentials are required they may be provided by duplicating the whole system the required number of times. Alternatively the highest required potential can be supplied and lower potentials derived from it via monolithic regulators. This method becomes inefficient where the expected current demand at the highest potential is low and ineffective where a low current source at a negative potential is required. In these cases a converter system becomes more appropriate. Frequently it may be more convenient to produce the derived supply in the equipment near to where it is required rather than incorporating it in the power supply unit. The system diagram then may be drawn as shown in Figure 33.

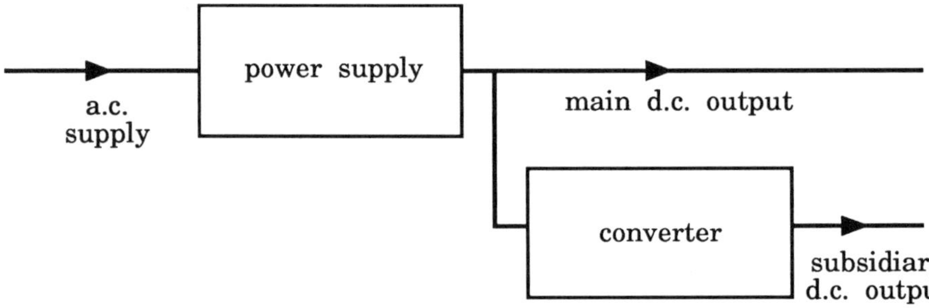


**Figure 33. The subsidiary power supply system.**

*Low power converter systems.*

Important applications of power converters occur in computer logic systems which typically have a +5 V supply but require a negative supply at low current for analogue to digital converters and supplies at +12 V and –12 V for RS 232 data transfer systems.

The power conversion process for the production of a negative supply proceeds in two stages, the production of a series of (positive) d.c. pulses which are then passed to a potential "level shifter" as shown in the system diagram in Figure 34.

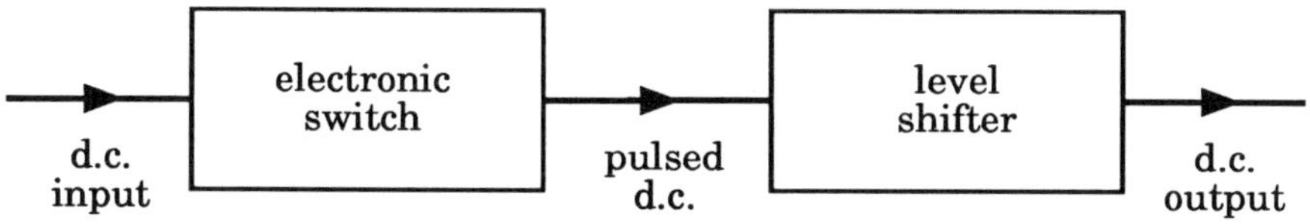


**Figure 34. The low current potential converter system.**

The electronic switch may typically consist of an astable system, acting as a pulse generator, controlling a transistor, darlington transistor or FET acting as a power switch. The realisation of the astable might use logic gates, the 555 timer i.c. or discrete components. Some form of buffer or power amplifier may be required for successful operation of the switching device.

One type of level shifter commonly employed makes use of two rectifier diodes and two capacitors as shown below.

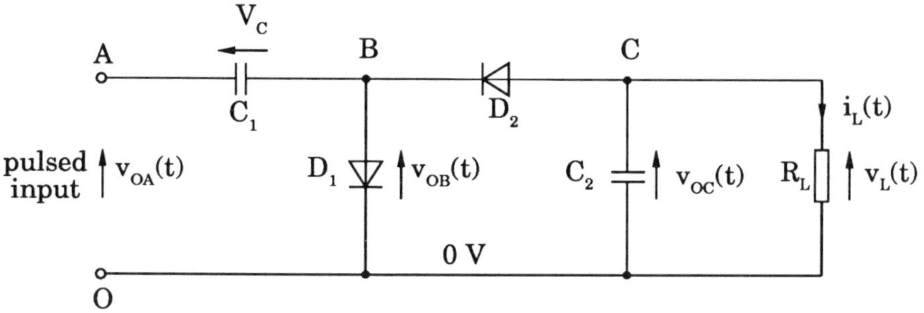


N.B. $v_L(t)$ and $i_L(t)$ are negative

**Figure 35. The level shifter.**

Analysis of the action of the circuit is simplified if it is assumed that the system has been running for sufficient time to reach a steady state and that the load current, $i_L(t)$, is zero. Suppose that the input, to point A, is a series of square pulses of amplitude 5 V.

The diode $D_1$ will prevent point B from rising above about 0.7 V (at which potential $D_1$ will conduct). The current drawn by the load is very small so given the short period of the pulses the charge on $C_1$ can be considered constant and therefore the p.d. across it, $V_c$,it will also be constant at about 4.3 V. Therefore, as the potential at A falls from +5 V to 0 V the potential at B will fall from +0.7 V to –4.3 V. The –4.3 V pulses appearing at B will cause $D_2$ to conduct until the potential across $C_2$, $v_{oc}$ , reaches about –3.6 V. $D_1$ will be reverse biased and will not conduct.

We see, then, that the maximum output p.d., $v_L$, will be about –3.6 V.

As soon as current is drawn by the load the output p.d. will fall and a ripple p.d. will appear on it. Where the pulse repetition frequency is above 10 kHz and capacitors of about 1 μF are used the maximum useful output current will be about 0.5 mA. Larger capacitors or a higher operating frequency will give higher outputs. At frequencies above about 50 kHz high speed switching diodes will be required for efficient conversion.

Graphs of the p.d.s to be found at points A,B and C on Figure 35 are shown below.

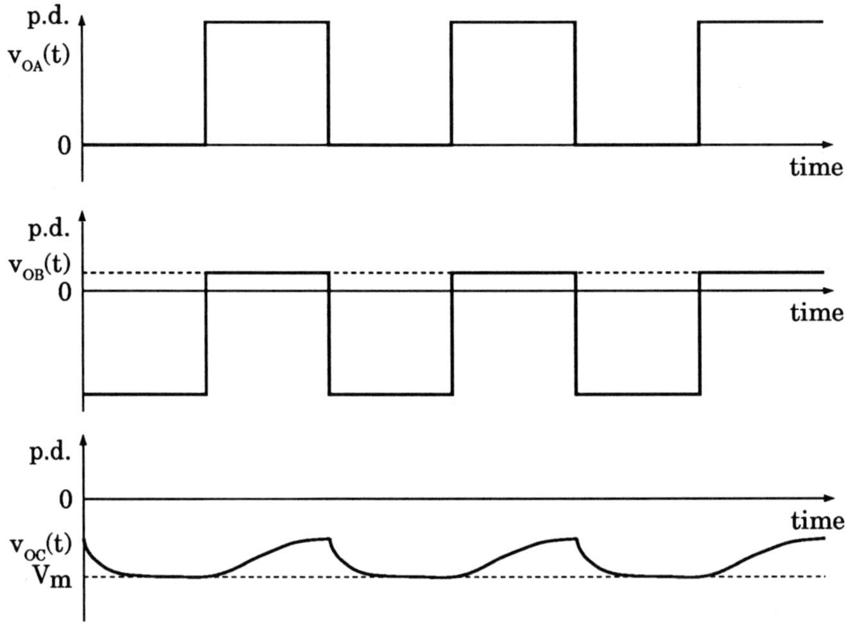


**Figure 36. Graphs of p.d.'s for Figure 35.**

Where larger p.d.s are required the circuit of Figure 34 can be cascaded as many times as required as shown in the circuit overleaf.

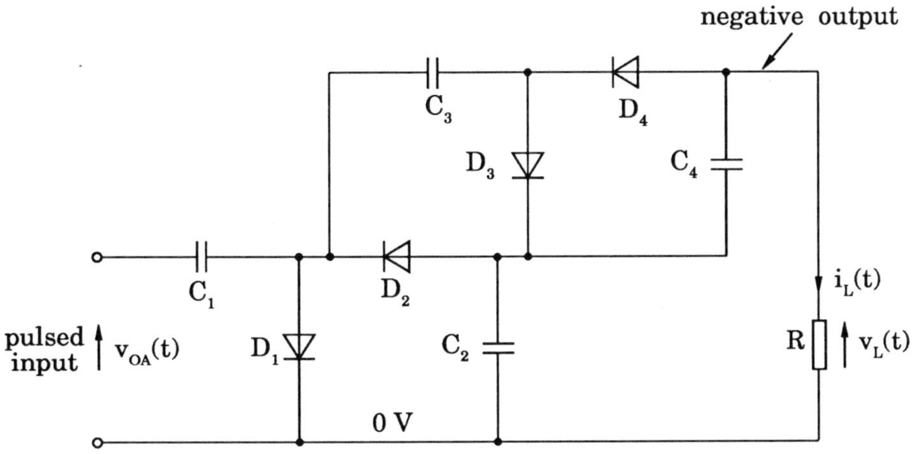


N.B. $v_L(t)$ and $i_L(t)$ are negative

**Figure 37. Two stage level shifter to give about –7.2 V from a +5 V pulsed input.**

A similar circuit can be used to produce increased positive potentials from a pulsed positive d.c. supply.

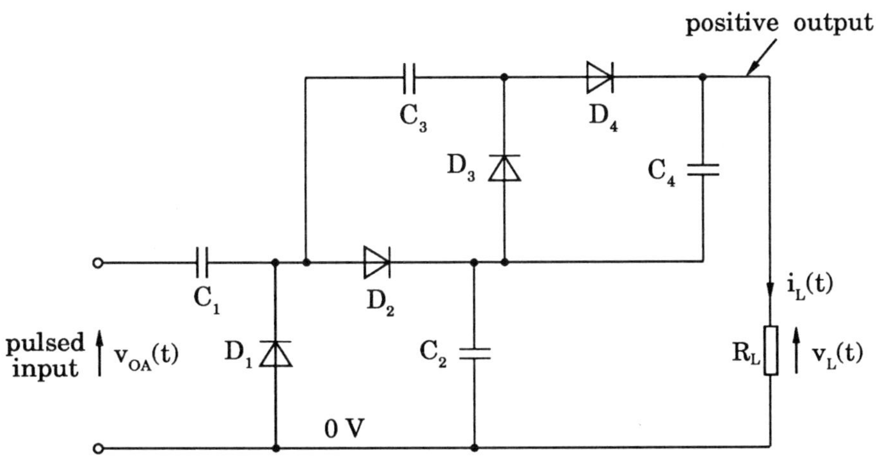


**Figure 38. Level shifter to produce a positive output.**

Drawing current from these converters immediately causes a drop in the output p.d. and the appearance of a ripple superimposed upon it. The output could be taken via one of the regulator circuits already discussed, but the current used by the regulator would be a significant proportion of that available from the converted supply.

A more subtle approach is to increase the value of the capacitors in the level shifter circuit, to decrease the ripple, and incorporate a feedback loop controlling the action of the pulse generator to improve regulation. The arrangement of the feedback loop is shown in Figure 39.

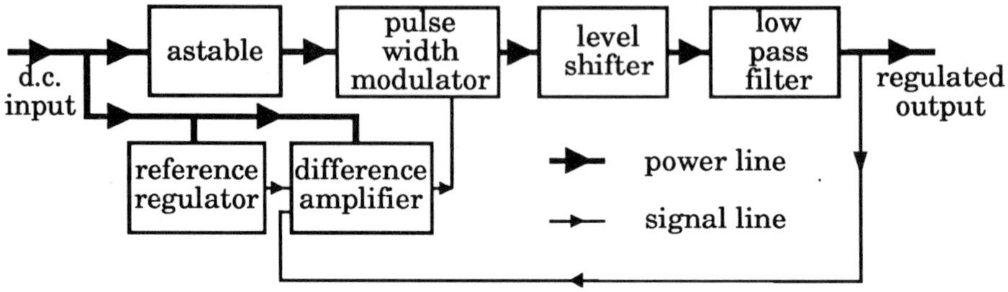


**Figure 39. The stabilised d.c. to d.c. converter system.**

Although it is possible to build a required d.c. to d.c. converter system from discrete components, for many applications an integrated circuit is readily available. Types 7660 and 7661 are suitable for supply potentials of 1.5 V to 10 V and 4.5 V and 20 V respectively from which they will produce a negative supply potential of the same magnitude as the input p.d. with an output impedance of around 100 Ω. Where an increase in the output p.d. is required or higher load currents are expected the converters may be cascaded or paralleled.

*High potential converters.*

Occasionally a low current source at high p.d. is needed for which the regulation requirement is not too stringent, for example for driving gas discharge displays or Geiger Muller tubes. In this case an expedient system utilises a power oscillator driving a transformer.

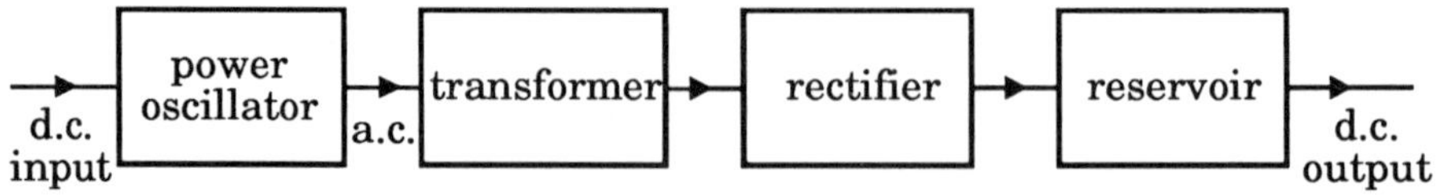


**Figure 40. An oscillator d.c to d.c. converter system.**

A very simple approach is to use a miniature 240 V mains transformer with a centre tapped low voltage secondary. The secondary is used as the primary (and feedback winding where required) and the mains winding becomes the secondary in this application.

The a.c. output from the transformer is rectified, smoothed and filtered in the ways already discussed. Two possible circuit configurations are shown in Figure 41.

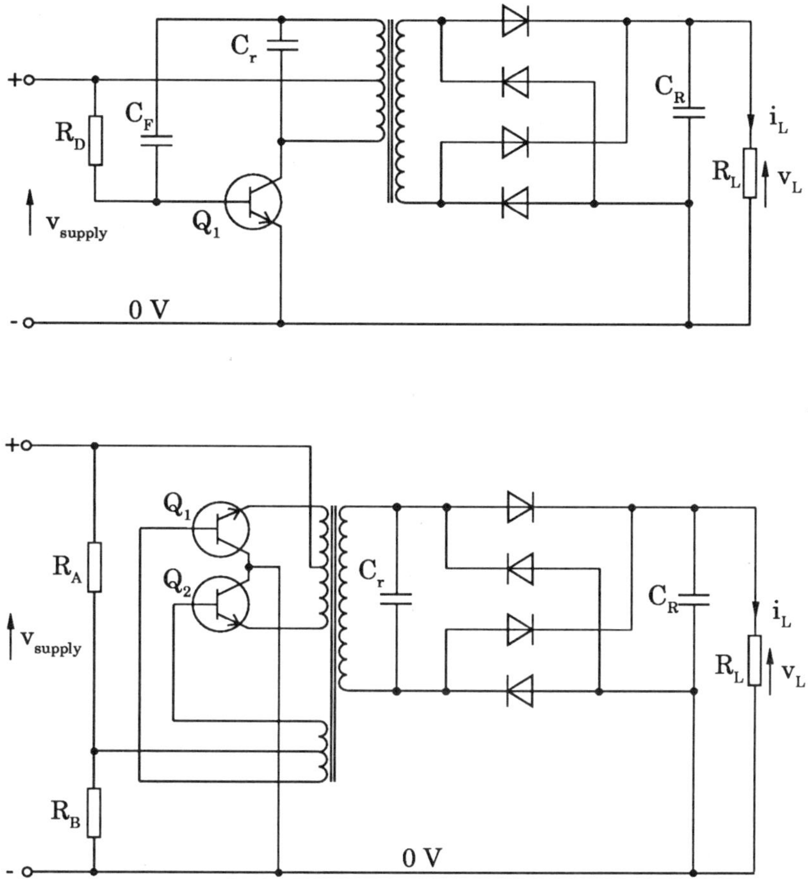


**Figure 41. Two oscillator/transformer d.c. to d.c. converter circuits.**

The first circuit shown above uses a Hartley oscillator which will tend to produce fewer high power (possibly troublesome) high frequency harmonics than the second circuit which has better power handling capabilities. The latter circuit is sometimes referred to as a multivibrator arrangement. Simple circuits such as these could be made to operate at high power levels but they are inherently poorly regulated. Conversion efficiencies of about 80% are achievable with these systems.

**NOTE**

**High voltages generated by converter systems are equally as lethal as those from the mains supply!**

*High power converter systems.*

Where large changes of potential are involved and regulation of the output is important the simple oscillator/transformer system may not meet requirements. If loads may draw currents in excess of a few milliamps, or large changes of potential are involved, then the use of regulators to reduce potentials or level shifters to increase potentials becomes inappropriate. If an increase of output p.d. is required at high power, for example in equipment intended to be powered from a 12 V vehicle supply, a suitable converter may employ some form of power switching system in which pulsed d.c. is fed to a transformer. The output from the transformer being rectified and filtered in the usual way. Although apparently similar in effect to the circuits of Figure 40 this system, as outlined in Figure 42, has the advantage of very efficient regulation of the output.

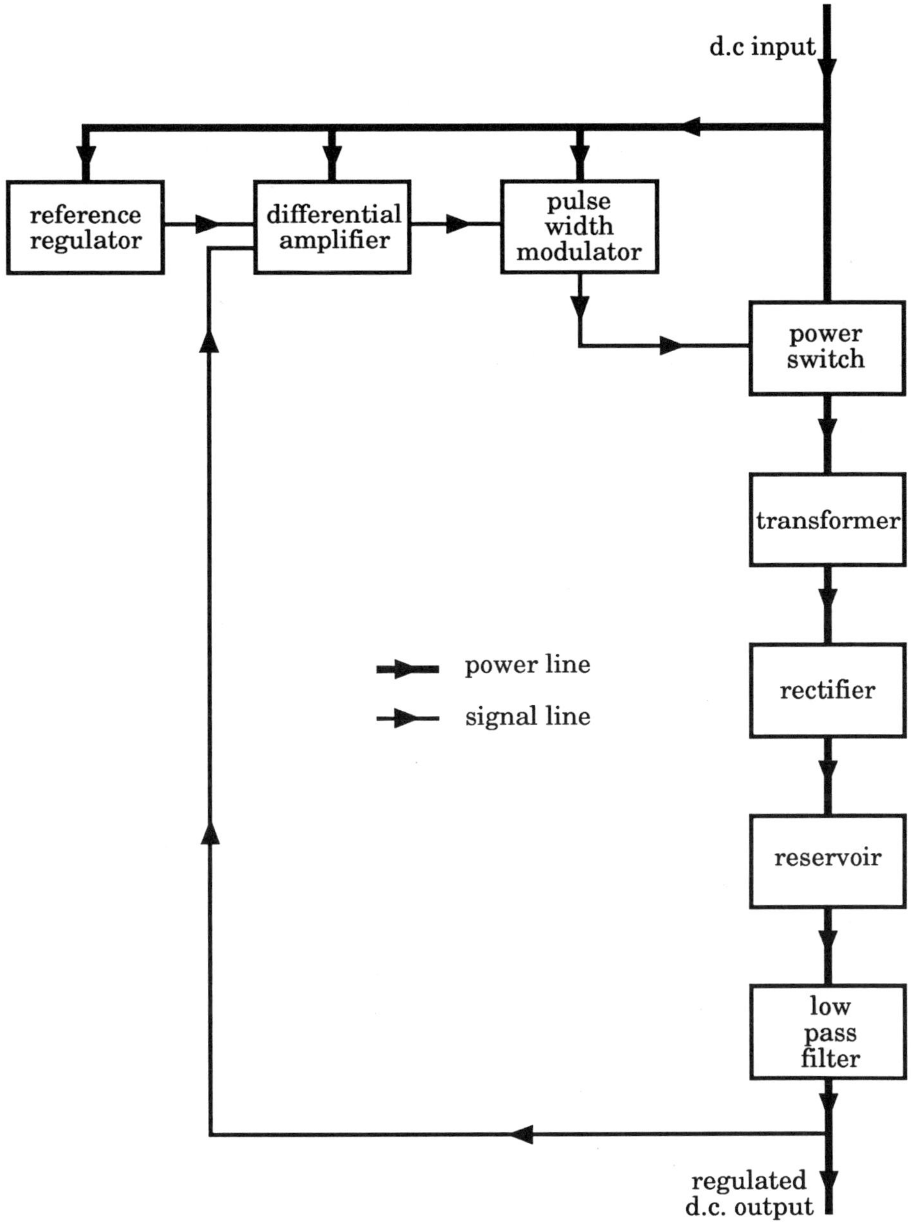


**Figure 42. A basic switching regulator system.**

# APPENDIX

*Photo-Voltaic (Solar Cell) Power Supplies - an example project.*

The applications of solar cells as a power source for space probes and earth satellites, and as a so-called renewable energy source, have generated popular interest. The application of solar cells may be used as a stimulus for the study of power supply systems in general. The discussion which follows gives a brief technical background of a project in which a solar cell system is developed. No claim is made that this project achieves perfection, merely that it fulfils the specification.

*Specification*

The system is to maintain the charge in a small 9 V Nickel Cadmium rechargeable battery (nicad) which is used to power a transistor radio for short widely dispersed periods.

*Development*

At the start of a project it is advisable to draw up a system diagram incorporating the functions which appear to be required in order to meet the specification. As development of the project continues this system diagram may be modified in the light of acquired knowledge. A suitable starting point based on Figure 34.

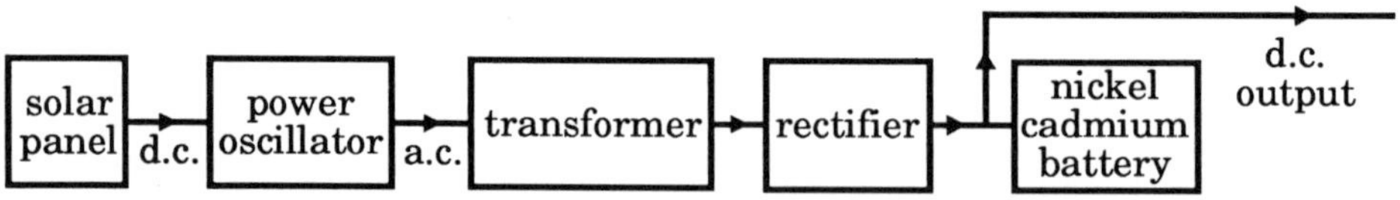


**Figure A1. A solar cell power supply system.**

The second stage of the project involves research about the various devices and circuits available for the implementation of the sub-systems.

The characteristics of suitable photo-voltaic cells might first be examined. At the time of writing silicon photo-voltaic cells approximately 90 mm x 60 mm are readily available at moderate cost. The output characteristics of a typical cell were measured in full sunlight and the results are shown graphically in Figure A2.

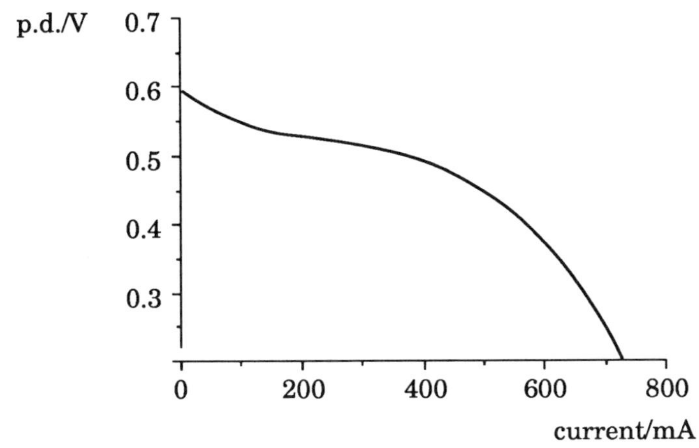


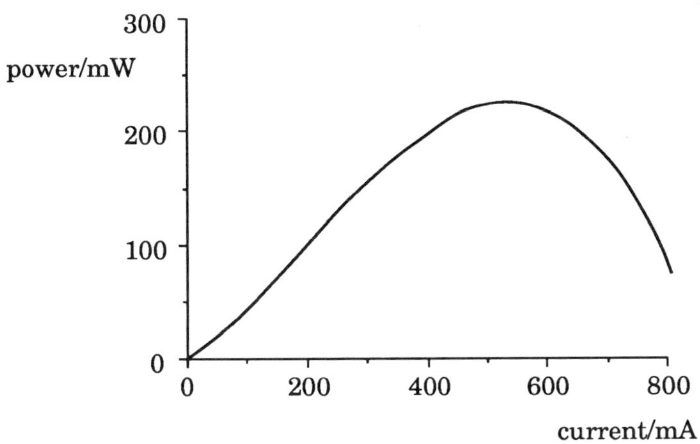


**Figure A2. Output characteristics of a typical solar cell in sunlight.**

It can be seen that the most efficient use of these cells occurs when the load draws approximately 0.5 A at 0.45 V and some 0.225 W are available. If the higher voltage required by the nicad and the electronic system is to be made available then a converter system must be employed. Suitable converter devices can now be assessed.

In previous sections of this text, converter systems using bipolar transistor circuitry were described. Bipolar devices which are suitable for switching supplies as low as 0.5 V are not available. Devices able to switch low voltages at low impedance are required.

The characteristics of two N-channel power MOSFETS are shown below.

| Device | $R_{DS}$ (Ω) | $I_D$ (A) | $P_D$ (W) |
|---|---|---|---|
| SMP50N06 | 0.028 | 50 | 125 |
| SMP60N06 | 0.023 | 60 | 125 |

The data reveal that these devices, when switched on, present a very low resistance. The power dissipations anticipated are well within the switching capabilities of these devices, with no heat sinks fitted. A glance at the transfer characteristics of power MOSFETs reveals, however, that the devices will only be switched fully ON (saturated) when $V_B$ exceeds about 4.5 V, although virtually no current will be used to control the switching.

Here we have a dilemma: power MOSFETs will provide a solid state switch with the desired characteristics, but they cannot be controlled using the p.d. available from the solar cell. A solution is to power the oscillator and control system from the rechargeable battery which is then recharged by the converted low voltage supply from the solar cells.

It seems sensible to store as much energy as possible, subject to the safe operating conditions of the battery, and to limit the use of energy from the battery when sufficient light to permit operation of the converter is not available. These criteria enable a final system diagram to be drawn up as shown in Figure A3.

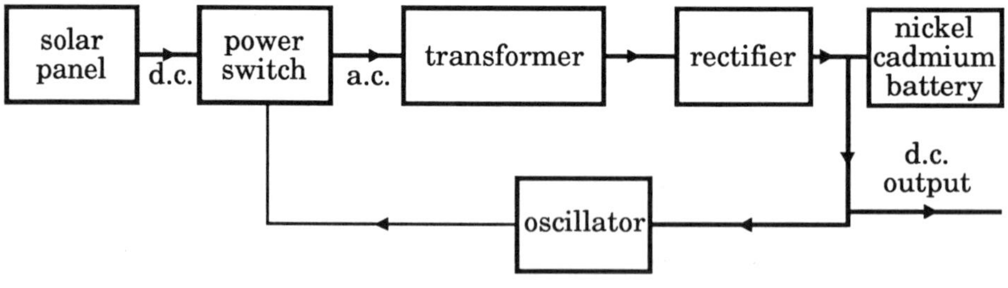


**Figure A3. The final solar power supply system diagram.**

A circuit diagram can now be envisaged which incorporates the devices considered appropriate. Ideally as many alternative realisations as possible for each part of the system should be considered. Each having been tested, the most appropriate would be selected and a circuit design finalised.

A simple example of such a circuit has been produced and is shown in Figure A4.

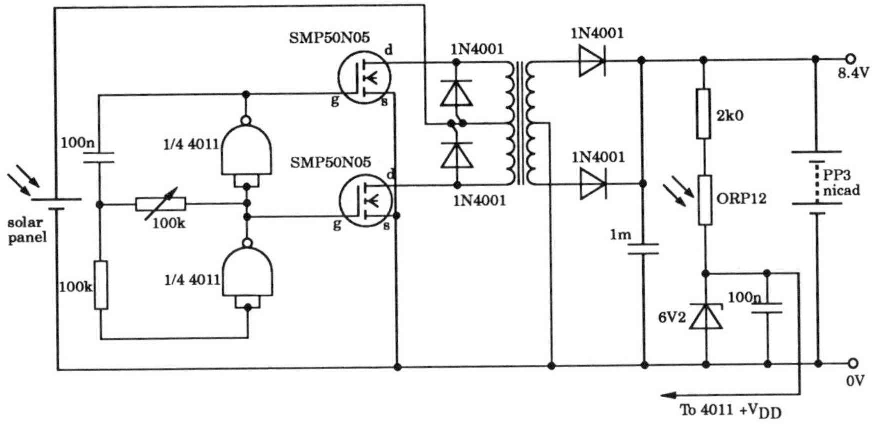


**Figure A4. A solar power supply circuit diagram.**

The oscillator circuit utilises digital logic gates in a relaxation oscillator. If CMOS gates are used power consumption can be minimised, and the output conveniently matched to the requirements of the power MOSFETs. Two, out of phase, signals can be made available via a push-pull arrangement shown in Figure A4, ensuring the most efficient use of the available energy.

It is possible to wind suitable transformers using commercially available ferrite core and coil former kits. If a coil winding machine is not available it may be more convenient to utilise ready wound components. Low voltage mains transformers used backwards, i.e. with the switched low voltage d.c. applied to the low voltage (originally the secondary) windings, form efficient step up transformers. The low powers involved may suggest the use of small transformers; unfortunately such components tend to have **high** d.c. resistances. The use of relatively large components will confer greater efficiency. In a prototype a transformer with two 6 V @ 4 A windings was used.

A very simple way of maximising energy storage and minimising energy loss is to include a Light Dependent Resistor (LDR) in the supply from the battery to the oscillator i.c. The extremely high dark resistance limits the current supplied to the oscillator to the point where oscillation ceases and only the CMOS standing current (around 1 μA) is drawn from the supply. The zener regulator provides simple stabilisation of the supply to the oscillator, the zener voltage being chosen to provide efficient switching of the MOSFETs.

The third stage in the development of the project involves the construction of a prototype, testing it and incorporating modifications until the desired specification is obtained. Finally a production layout and construction might be designed and fabricated.

N.B. Although this prototype functions, and meets the limited specification drawn up, it is not suggested that it represents the best practical arrangement possible. It would be possible, for example, to incorporate a charge control function in the oscillator sub-system.

# INDEX

NEMEC